APPLIED MISSING DATA ANALYSIS IN HEALTH SCIENCES

WILEY SERIES IN STATISTICS IN PRACTICE

Statistics in Practice is an important international series of texts which provide detailed coverage of statistical concepts, methods, and worked case studies in specific fields of investigation and study.

With sound motivation and many worked practical examples, the books show in down-to-earth terms how to select and use an appropriate range of statistical techniques in a particular practical field within each title's special topic area.

The books provide statistical support for professionals and research workers across a range of employment fields and research environments. Subject areas covered include medicine and pharmaceutics; industry, finance, and commerce; public services; the earth and environmental sciences, and so on.

The books also provide support to students studying statistical courses applied to the above areas. The demand for graduates to be equipped for the work environment has led to such courses becoming increasingly prevalent at universities and colleges.

It is our aim to present judiciously chosen and well-written workbooks to meet everyday practical needs. Feedback of views from readers will be most valuable to monitor the success of this aim.

A complete list of titles in this series appears at the end of the volume.

APPLIED MISSING DATA ANALYSIS IN HEALTH SCIENCES

XIAO-HUA ZHOU
University of Washington

CHUAN ZHOU
University of Washington

DANPING LIU
National Institutes of Health

XIAOBO DING
Chinese Academy of Sciences

WILEY

Published by John Wiley & Sons, Inc., Hoboken, New Jersey.
Published simultaneously in Canada.

Library of Congress Cataloging-in-Publication Data:

Zhou, Xiao-Hua, author.
Applied missing data analysis in the health sciences / Xiao-Hua Zhou, Chuan Zhou, Danping Liu, Xiaobo Ding.
p. ; cm.
Includes bibliographical references and index.
ISBN 978-0-470-52381-0 (cloth)
I. Zhou, Chuan, 1972- author. II. Liu, Danping, 1981- author. III. Ding, Xiaobo, author. IV. Title.
[DNLM: 1. Data Interpretation, Statistical. 2. Biomedical Research–methods. 3. Models, Statistical. 4. Research Design. WA 950]
R853.S7
610.72′4–dc23

2013047830

Printed in the United States of America

10 9 8 7 6 5 4 3 2 1

To Yea-Jae, Yi,
Tingting, and Shuqin

CONTENTS IN BRIEF

CONTENTS

LIST OF FIGURES

LIST OF TABLES

PREFACE

With a strong practical emphasis on health science applications, this book describes statistical methods and models for the analysis of data with missing values. We attempt to write so that researchers with experience in applied data analysis, but less technical knowledge than a statistician, should be able to understand and implement most of the methods described. For those with a stronger background in statistics, we provide more technical details as to not detract from the flow of rest of the chapter. We have also tried to choose examples that are relevant to most health science researchers who work in a variety of disciplines.

In all fields of study, missing data are a common problem since, for any data collection process, there are so many things that could go wrong that missing values are all too likely. Thus, when attempts are made to answer the scientific questions of interest, researchers ask the all-too-common question: what do we do with the missing data?

The statistical literature to answer this question is well developed, but overly technical and complicated for researchers who are not experts in statistics and methodology. Therefore, researchers may recognize the existence of missing data, but fail to respond for two reasons: first, they may not understand the consequences of ignoring missing data and how it can impact the validity of their results; second, there is a lack of understanding of the statistical methods for missing data and how to apply them in their own research. Therefore, the purpose of this book is to provide health science researchers with the means of understanding the importance of missing data in their own personal research and the ability to use these methods given the available software.

This book is organized into eight chapters. Chapter 1 introduces concepts on the missing data mechanism and some real-world examples. Chapter 2 gives an overview of methods for dealing with missing data. Chapter 3 describes some design strategies for minimizing the impact of missing data. Chapters 4 and 5 introduce methods for dealing with missing data problems in cross-sectional and longitudinal studies, respectively. Chapter 6 deals exclusively with missing data problems in survival analysis. Whereas Chapters 3–6 deal with ignorable missing data problems, Chapter 7 presents methods for dealing with nonignorable missing data problems. Finally, Chapter 8 discusses methods for dealing with missing data in causal inferences.

As we worked through examples in the book, we chose to provide software code in the text of the chapters as we want to encourage application of these methods after an understanding of the basic theory. We chose to include R code in the text as many of the methods can be implemented in R; in addition, R is also a publicly available software environment (see www.r-project.org). Since many researchers also use Stata in addition to R, we include code for some selected examples. All the analysis data sets, together with R and Stata codes used in this book, can be downloaded from http://faculty.washington.edu/azhou/.

X.H. Zhou, C. Zhou, D. Liu, and X. Ding

Seattle, Washington
March, 2014

CHAPTER 1

MISSING DATA CONCEPTS AND MOTIVATING EXAMPLES

1.1 OVERVIEW OF THE MISSING DATA PROBLEM

Data are the fundamental building blocks of valid statistical inference for biomedical and social sciences research. Unfortunately, for many reasons, more often than not we will be missing some observations. Data are sometimes missing by design, such as in two-stage case-cohort designs. There are situations when missing data are not relevant to the analysis and therefore can be safely ignored. So, it is important to understand what we mean by missing data in this book. According to Little et al. (2012b), missing data are defined as values that are not available, but otherwise would be meaningful for analysis if they were observed. Even in the case of missing data, the goal remains to make inferences about the population targeted by the complete sample. Unfortunately, there is no universal method for handling a missing data problem. This is because the selection of subjects for a study is usually known, but the process by which observations on those subjects become missing–the missingness mechanism–is usually unknown, and the data alone cannot definitively inform us about this process. Therefore, with missing data, additional assumptions are required in order to proceed with analysis, and the validity of these assumptions cannot be determined from the observed data alone. For this reason, assessing the sensitivity of conclusions to the assumptions should play a central role in any analysis of data with missing values. In fact, any analysis should principally include the hypothesis under investigation, the information on the observed data, and the reasons for the missing data. When data

Applied Missing Data Analysis in Health Sciences, First Edition. Xiao-Hua Zhou, Chuan Zhou, Danping Liu, and Xiaobo Ding.

are missing, information is lost, and the value of what remains depends on whether we can identify plausible reasons for missing the data and on the sensitivity of the study conclusions to different assumptions on missingness mechanisms.

Over the years, our view on missing data problems has evolved substantially as we have obtained new insights into the problem and learned to deal with it. In the beginning, missing values were merely an inconvenience as holes in a data matrix that cause statistical software to crash. Therefore, early work on the problem was largely on computation (Afifi and Elashoff, 1966; Dempster et al., 1977). Later we realized that even if data were not there, information could still be obtained from the way data were missing or from the relationship between the missing and observed values. This led to the missingness mechanism categorization and imputation based on auxiliary variables. We found that augmenting the observed data with the extracted additional information, along with appropriate inference adjustment, can improve the statistical analysis (Little and Rubin, 2002; Schafer, 1997b). Later researchers further realized that missing values are part of a more general concept of *coarsened data*, which include numbers that have been grouped, aggregated, rounded, censored, or truncated, resulting in partial loss of information (Heitjan and Rubin, 1991). Missing data methods may be extended to a broader research on analysis of partially observed data that encompassed a wide range of advanced statistical topics such as double robustness, causal inference, and semiparametric theory (Robins and Rotnitzky, 1995; Tsiatis, 2006).

The statistical literature on the missing data problem is now quite extensive with many excellent textbooks available. Little and Rubin (2002) provided a good overview of likelihood methods and an introduction to multiple imputation. Allison (2001) presented a less technical overview intended for technically less sophisticated audience. Schafer (1997b) covered the EM algorithm and data augmentation for multivariate normal and general location models from a Bayesian perspective. Molenberghs and Kenward (2007) focused on intervention-based clinical studies, whereas Daniels and Hogan (2008) focused on longitudinal studies with a Bayesian emphasis. Two recent textbooks provide excellent discussion of a particular statistical method for analyzing and drawing inferences from incomplete data, namely, multiple imputation (MI) (Carpenter and Kenward, 2013; van Buuren, 2013).

In this book, we are trying to strike a balance between theory and application. This book is intended for researchers and graduate students in the biomedical fields who are moderately sophisticated in terms of statistical skills. We cover a wide range of statistical methods for dealing with missing data problems for those who are interested in theories. For those who wish to analyze incomplete data, we provide detailed working examples, using mostly freely available statistical software R (R Core Team, 2013) or the commercial software Stata, although many approaches are readily available in other statistical software packages.

1.2 PATTERNS AND MECHANISMS OF MISSING DATA

Before the explosion of data in the modern age of the Internet and informatics and before the age of clinical trials, data mainly came from agriculture and survey studies.

Such studies measured all variables of interest for each unit (or case, observation, subject, depending on the context). Measured variables may have included continuous variables (e.g., weight and height), categorical variables (e.g., gender, race, and level of education), and/or repeated measures on the same variable over time. Such data are often organized in a rectangular format, where the rows of the data matrix represent the units or subjects and the columns represent variables measured for each unit. Since statistical methods developed for the missing data setting refer to the necessary *shape* of the data, we will continuously refer to the data structure, or the data matrix, of the study as well.

In the initial stages of development, most missing data techniques were developed specifically for the missing data in this type of study with data in a rectangular form; see, for example, the seminal work of Little and Rubin (2002). But missing data may also occur in studies where, by design, not all variables are measured on each subject, or where subjects are not measured at the same time points, giving this type of data a different shape. For example, in studies involving survival following transplantation, patients often have unequal length of follow-up and a different number of follow-up visits. So for today's ever-growing complex study designs, special efforts are required to reshape the data before we can apply the missing data techniques.

Central to the analysis of incomplete data is an understanding of why the data are missing and its implications for the analysis. In the context of the missing data problem, it is necessary to distinguish between *patterns* of missing data and *mechanisms* of missing data. Missing data patterns describe which values are observed in the *data matrix* and which values are missing, whereas the missing data mechanism describes how the process underlying the missingness is related to the values in the data matrix. As the readers will see later, certain missing data patterns such as a monotone pattern may lead to more efficient algorithms for imputation. On the other hand, many statistical techniques rely on correct assumptions regarding the missing data mechanisms, which can sometimes be inferred by examining missing data patterns.

1.2.1 Missing Data Patterns

We illustrate typical missing data patterns using some examples. The first four examples come from the R package **mice** (van Buuren and Groothuis-Oudshoorn, 2011).

Example 1.1 (univariate missing data pattern). We illustrate the univariate missing data pattern with the following data named ''pattern1'' in the R package mice.

```
   A  B  C
1 88 42 63
2 84 82 12
3 26 59 66
4  7 28 73
5 12 75  2
6 79 41 NA
7 81 84 NA
8 64 19 NA
```

This data set has three variables A, B, and C, and the variable C has missing values, denoted by ''NA''. It can be seen that five of the eight subjects in the data set are completely observed, whereas the values of variable C for the other three subjects are missing. This is the simplest missing data pattern. More complex patterns are illustrated in the following examples.

Example 1.2 (monotone missing data pattern). Attrition (dropout) in longitudinal studies is a very frequent occurrence, especially for long follow-ups or difficult study populations such as children. If in a study subjects who drop out before the end of the study do not return, then this scenario leads to the so-called monotone missing data pattern.

```
    A   B    C
9  60  11   60
10 91  85  100
11 20  95   96
12 95  37   NA
13  1  26   NA
14 61  26   NA
15 18  NA   NA
16  7  NA   NA
```

The above data set, named ''pattern2'' in the R package mice, is an example of the monotone missing data pattern. It can be seen that if a subject does not have its variable B observed, then accordingly its variable C is not observed too.

Example 1.3 (file-matching missing data pattern). Sometimes two or more variables are collected on units, but are never jointly observed. In the following data set named ''pattern3'' in the R package mice, we see that if variable B is observed, then the corresponding variable C is not, and vice versa.

```
    A   B   C
17 45  34  NA
18 99   6  NA
19 16  84  NA
20 79  37  NA
21 38  95  NA
22 32  NA   5
23 48  NA  34
24 91  NA  25
```

It is important to know that for this missing data pattern, parameters relating only to the association between these jointly unobservable variables are not estimable from the data. In the above data, there is no information in the data about the partial associations of B and C given A. In practice, analyses of data with this pattern require to make strong assumptions about these partial associations.

Example 1.4 (general missing data pattern). Consider the situation of unit nonresponse and item nonresponse in a survey study. Unit nonresponse occurs when a questionnaire is administered and a subset of subjects do not complete the questionnaire because of noncontact, refusal, or some other reasons. Item nonresponse occurs when some subjects complete the questionnaire but do not answer all items. Such missing responses result in a data matrix with holes in it without a particular pattern.

```
     A   B   C
25  26  88  32
26  42  66  21
27  86  54  NA
28   9  92  NA
29  20  83  NA
30  89  NA  41
31  NA  NA  35
32  NA  NA  33
```

A data set named ''pattern4'' in the R package mice is shown above. It is seen that there is no particular pattern for this data set.

Example 1.5 (causal effect) Under Rubin's causal model and potential outcome framework (Rubin, 1974), let $Y_t(u)$ denote the outcome of subject u under treatment t and $Y_c(u)$ denote the outcome under treatment c for the same unit, then the *causal effect* of the cause t on unit u as measured by Y and relative to cause c is the difference between $Y_t(u)$ and $Y_c(u)$. In a randomized controlled trial, if a subject is assigned to the intervention arm, then his outcome had he been assigned to the control arm is automatically unobserved, and vice versa. These potential outcomes may be treated as missing data and handled using methods described in this book. The missing data pattern is the same as that in the above file-matching problem since the two potential outcomes can never be observed simultaneously on the same unit.

1.2.2 Missing Data Mechanisms

Understanding the missing data mechanism is important for the analysis of the missing data. The rationale of using a method for dealing with a missing data set relies on the missingness mechanism. To explain the missing data mechanisms, we first define a vector of indicator variables R that identify which variables are observed and which are missing. For example, $R = (0, 1)^\top$ indicates that the first variable is missing and the second variable is observed. There are two underlying requirements for missing data indicators. First, missingness indicators contain information about the missing values. Clearly, they are useless otherwise. Second, the values that indicators hide must be meaningful. Sometimes the missingness indicators hide undefined values and the problem should not be considered a missing data problem.

The missing data mechanism describes how the missingness indicator vector R is related with the vector of study variables Y, or in other words, the conditional distribution of R given Y, say $f(R|Y)$. Here, we regard R as a random vector, which formalizes the role of the missing data mechanism in modern missing data analysis. It is best to view the distribution of R as a mathematical device to describe the rates and patterns of missing values and to capture roughly possible relationships between the missingness and the values of the missing items themselves (Schafer and Graham, 2002).

Since some of the variables Y may be missing, we partition Y into two parts, Y_{obs} and Y_{mis}, the observed part and the missing part, respectively. Below we introduce the three missing data mechanisms: missing completely at random (MCAR), missing at random (MAR), and missing not at random (MNAR).

The data are called MCAR if missingness does not depend on the value of data Y, missing or observed. Statistically,

$$f(R|Y) = f(R) \text{ for all } Y. \tag{1.1}$$

It means that the missingness indicator is conditionally independent of the study variables, so data on the study variables, regardless of missing or observed, have nothing to do with the mechanism that the data are missing. Consider an experiment involving random sampling. Under MCAR, people with missing values are really just a random sample of the study population. As a result, people with observed data still make up a valid random sample from the study population for the ultimate inference, though with some loss in efficiency because of the smaller sample size. Therefore, when data are MCAR, we can ignore the units with missing values and proceed with the complete-data analysis, and still make valid inferences.

MAR is a less restrictive assumption on the missing data mechanism than MCAR. Data are called MAR when missingness depends only on the observed components Y_{obs} of Y, but not on the missing components Y_{mis}. Statistically,

$$f(R|Y) = f(R|Y_{\text{obs}}) \text{ for all } Y_{\text{mis}}. \tag{1.2}$$

Despite its name, MAR does not mean that the missing data are a simple random sample of all data values, which is the case for MCAR. MAR lessens MCAR in that the missing values can behave like a random sample of all values within subpopulations defined by the observed data rather than the entire data values.

The mechanism is called missing not at random if the distribution of R depends on the missing values in Y. Strictly speaking, MNAR comes in two forms: (i) missingness depends on the missing value itself or (ii) missingness depends on an unobserved variable. The latter is common in latent variable models or hidden Markov models.

Notably, the classification has been criticized because of the confusing nature of its terminology (Schafer and Graham, 2002). We encourage readers to focus more on the implications than on the names. Practically, MNAR can be difficult to distinguish from MAR owing to the lack of information from missing values. This in part explains the popularity of the MAR assumption in the development of missing data techniques. But the situation can become really complex if a multivariate data set is considered, even more so if the data set contains variables with missing observations, which are MCAR, MAR, or MNAR.

We demonstrate the three mechanisms and their consequences through the following simple example.

Example 1.6 (missing data mechanisms) Consider a simple linear relationship between two variables x and y, where the predictor variable x is fully observed, but some of y are missing according to different missing mechanisms. The variables x and y follow the linear regression $y = x + \varepsilon$, where x and ε independently follow standard normal distributions. We generate an independent sample of (x, y) with size 300. The sample is referred to as the full data because all values of x and y are known. Then three samples with y-missing data are generated from the full data, according to the missing mechanisms MCAR, MAR, and MNAR, respectively. In the first sample, we randomly select a subset of y sample to be missing. Since the missingness of y does not depend on any values of x and y, the missing mechanism is MCAR. In the second sample, y is missing if $x < 0$, that is, whether y is missing depends only on the observed x. By the definition of MAR, the missing mechanism is MAR. In the third sample, y is MNAR by assuming that the missingness depends also on the unobserved y. Specifically, y is missing if $x < 0$ and $y < 0$.

The scatter plots of the three samples are presented in Figure 1.1a–c, respectively. We conduct the linear regression analysis based on three subsets of the samples: the full data, the complete-case data (i.e., only containing subjects with y observed), and the missing-case data (i.e., only containing subjects with y missing and it is assumed that the unknown y values are known to us). The regression lines are presented in Figure 1.1, and the regression results for the full data and the complete-case data are summarized in Table 1.1. It can be seen that under MAR or MNAR, results based only on the complete-case data may be biased.

1.3 DATA EXAMPLES

To illustrate the different analytic approaches and to assist the readers in applying the methods described in the book, we use data from a wide range of clinical studies, including cross-sectional data, longitudinal data, and survival data from both

TABLE 1.1 Regression Results Under Different Missing Data Mechanisms, the Estimates with Standard Deviations of the Intercept and the Coefficients of x, the R-Squared and Adjusted R-Squared, and the Number of Observations

	Full Data	MCAR	MAR	MNAR
(Intercept)	0.08 (0.06)	0.01 (0.08)	−0.07 (0.13)	0.44 (0.09)
x	1.02 (0.06)	1.01 (0.08)	1.15 (0.12)	0.70 (0.09)
R^2	0.53	0.54	0.36	0.23
Adjusted R^2	0.53	0.53	0.35	0.23
Number of observations	300	148	163	201

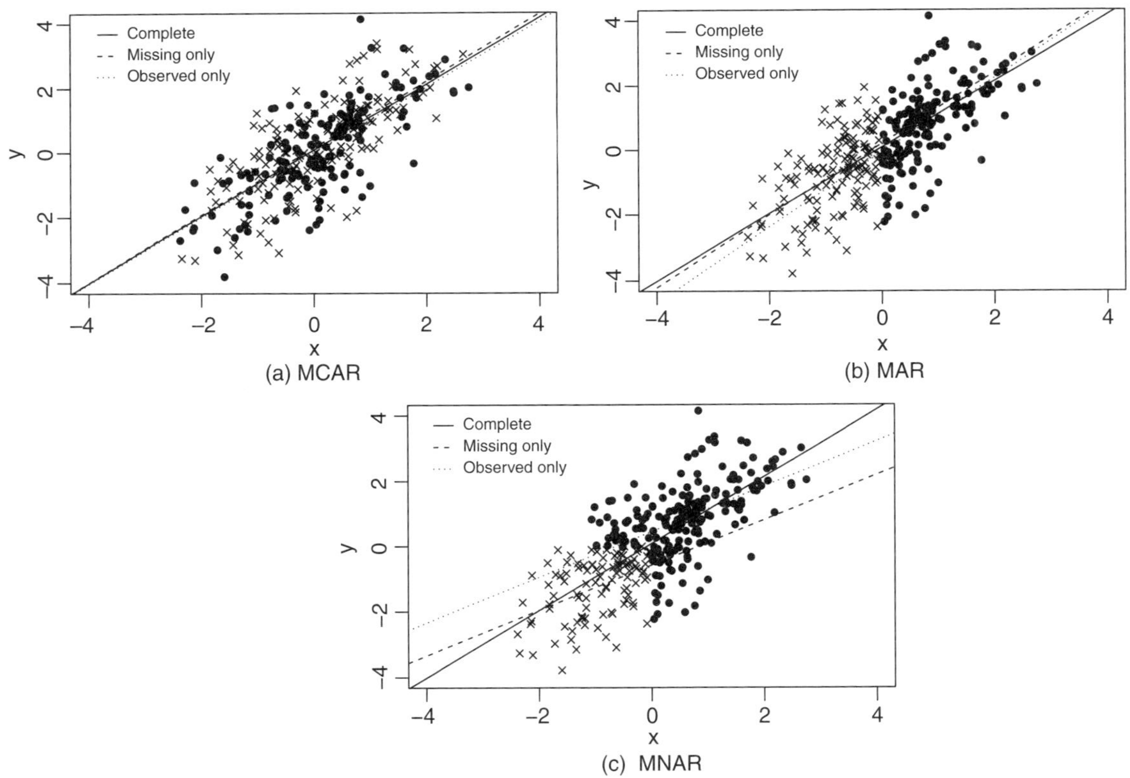

FIGURE 1.1 The solid circles are the observed sample points, and the x symbols are sample points with y values missing. The solid line is the regression line of y on x estimated from the full data, while the dotted and dashed regression lines are estimated from the subjects with completely observed and missing values, respectively.

randomized clinical trials and observational cohort studies. In the following, we provide some detailed description of the studies and briefly outline some of the missing data problems. Detailed missing data patterns will be examined in the subsequent chapters.

1.3.1 Improving Mood and Promoting Access to Collaborative Treatment (IMPACT) Study

1.3.1.1 Background In one of the largest treatment trials for depression in adults to date, a team of researchers supported by the Hartford Foundation followed 1801 depressed, older adults from 18 diverse primary care clinics across the United States from 1998 to 2003 (Unützer et al., 2002). The 18 participating clinics were associated with eight health care organizations in the states of Washington, California, Texas, Indiana, and North Carolina. The clinics included diverse health care systems: several Health Maintenance Organizations (HMOs), traditional fee-for-service clinics, an Independent Provider Association (IPA), an inner-city public health clinic, and two Veterans Administration (VA) clinics.

One half of the enrolled study participants were randomized to receive IMPACT collaborative care management, and the other half received the care normally available in their primary care clinics (including referral to specialty mental health care). The study examined the effects of the IMPACT intervention on depression, quality of life, costs, physical functioning, and other areas of physical and mental health.

IMPACT care is characterized by five essential elements: (i) it is collaborative care involving primary care physicians, a care manager, and a psychiatrist; (ii) the depression care manager plays a crucial role in coordinating and implementing the treatment plan; (iii) there is a designated psychiatrist; (iv) outcome measures are frequently assessed and monitored; and (v) stepped care allows us to adjust the treatment according to clinical outcomes.

Since the end of the trial, a number of organizations in the United States and abroad have adopted and implemented the IMPACT program with diverse patient populations. As reported in the December 11, 2002 issue of the *Journal of the American Medical Association*, the IMPACT model of depression care more than doubles the effectiveness of depression treatment for older adults in primary care settings.

At 12 months, about half of the patients receiving IMPACT care reported at least a 50% reduction in depressive symptoms, compared with only 19% of those in usual care. Analysis of data from the survey conducted 1 year after IMPACT shows that the benefits of the IMPACT intervention persist after 1 year. IMPACT patients experienced more than 100 additional depression-free days over a 2-year period than those treated with usual care.

1.3.1.2 Missing Data Problem For the IMPACT study, one of the primary outcomes is the depression as measured by the Symptom Checklist (SCL-20) scores assessed at 0 (baseline), 3, 6, 12, 18, and 24 months. Table 1.2 describes the missing data pattern of SCL-20 scores. Of the total subjects, 70% completed all six

TABLE 1.2 Missing Data Pattern for SCL-20 Scores Over Time

	Month							
Pattern	0	3	6	12	18	24	Number	Percent
1	✓	✓	✓	✓	✓	✓	1269	70
2	✓	✓	✓	✓	✓	·	60	3
3	✓	✓	✓	✓	·	·	60	3
4	✓	✓	✓	·	·	·	64	4
5	✓	✓	·	·	·	·	69	4
6	✓	·	·	·	·	·	117	6
7	?	?	?	?	?	?	162	10

assessments (pattern 1). A monotone missing data pattern was observed in 20% of the subjects (patterns 2–6), and the remaining 10% of subjects had intermittent missing data (pattern 7).

1.3.2 National Alzheimer's Coordinating Center Minimum Data Set

1.3.2.1 Background The National Alzheimer's Coordinating Center (NACC) was established by the National Institute of Aging (NIA, U01 AG016976) in 1999 to facilitate collaborative research. Collecting data from the 29 NIA-funded Alzheimer's Disease Centers (ADCs) across the United States, NACC has developed and maintained a large relational database of standardized clinical and neuropathological research data (Beekly et al., 2004, 2007). Two large databases are being maintained by the NACC study: one cross-sectional data collection (MDS) and one longitudinal data collection (UDS). More information about the NACC study can be found at the web site `http://www.alz.washington.edu`.

The NACC Minimum Data Set (MDS) consists of all subjects enrolled in the ADC's Clinical and Satellite Cores (1984–2005). The MDS contains one record for every subject enrolled and is analyzed as cross-sectional data. To demonstrate missing covariate(s) problem in a cross-sectional data analysis, we included a total of 17,403 deceased patients. The outcome variable is the mini-mental state examination (MMSE), which is a brief 30-point questionnaire test used to screen for cognitive impairment. The MMSE score can range from 0 to 30, with a lower score indicating a more severe impairment. We would like to examine the risk factors that affect the mean MMSE score, and examine whether the effects are modified by the patients' true Alzheimer's disease (AD) status. However, the gold standard ascertainment of AD, based on brain autopsy, is available only for about 31% of the sample. The missingness may be due to the patients' or their family's decision not to provide autopsy or unavailability of such a test. We believe that their decision of disease verification may be associated with the demographic characteristics (e.g., age, gender, and race), but it is unlikely to be correlated with their true AD status. So the MAR assumption seems to be reasonable. The risk factors we extracted from the database are age (continuous variable indicating age at the MMSE test), gender (binary variable with 1 indicating male gender), race (binary variable with 1 indicating white race), and marital status (binary variable with 1 indicating married); and clinical diagnosis of AD (binary variable with 1 indicating clinically diagnosed with AD), stroke (binary variable with 1 indicating having stroke before), Parkinson's disease (binary variable with 1 indicating presence of the disease), and depression (binary variable with 1 indicating presence of the disease). We included clinical diagnosis of AD as a risk factor and view it as a variable that captures patient history, clinical test results, and physician's clinical decision.

1.3.2.2 Missing Data Problem In this example, only one covariate variable (AD status) has missing values, whereas all other variables including outcome are completely observed. Among 17,403 subjects, 11,945 (68.6%) had the AD status missing, which is considered an important predictor for cognitive impairment.

1.3.3 National Alzheimer's Coordinating Center Uniform Data Set

1.3.3.1 Background The Uniform Data Set (UDS) at NACC is a multicenter longitudinal study that began in September 2005. Participants came from 29 ADCs across North America. The participants were referred or self-referred for evaluation of possible dementia, or recruited specifically for clinical research. Most subjects underwent clinical assessment and neuropsychological tests for cognitive impairment upon enrollment. Approximately once each year, the patients received periodical reevaluation and cognitive tests until their death or dropout from the study. Among these cognitive tests, the MMSE is a brief 30-point questionnaire test that is widely used to screen for cognitive impairment.

We used the same data set as in Monsell et al. (2012), who compared the decline of different measures of cognitive function in the NACC UDS data. Our focus is only on the MMSE score, and the aim is to investigate the risk factors affecting its rate of decline. Subjects satisfying all the following criteria were included in the analysis: (i) at least 65 years or older, (ii) with a clinical diagnosis of amnestic mild cognitive impairment (aMCI), and (iii) with at least one follow-up visit after the first diagnosis of aMCI. As a result, the analyzed data set had a total of 2830 subjects with 8752 visits. The covariates included age at the first diagnosis of aMCI, gender, clinical dementia rating sum of boxes (CDRSB), education (categorized as 0–12, 13–16, and 17+ years), Hachinski ischemic score (HIS) (categorized as 0, 1, and 2+ points), and apolipoprotein E (APOE) e4 alleles (yes/no). CDRSB and HI are time-varying covariates, whereas others are baseline covariates. More detailed background information could be found in Monsell et al. (2012).

1.3.3.2 Missing Data Problem In this example, missingness occurs in both subject and observation levels: 3.4% of the visits had missing HIS; 4.6% of the visits had missing MMSE; 33.5% subjects had missing APOE e4; and 0.1% subjects had missing education.

1.3.4 The Pathways Study

1.3.4.1 Background These data are used to illustrate missing data methods when dealing with missing covariates in survival analysis.

To study the association of depression with diabetes outcomes, investigators from the University of Washington and the Group Health Research Institute initially developed the Pathways study, which is a prospective, population-based, observational study (Lin et al., 2004). Subjects were recruited into the study in March 2001. Baseline surveys were conducted, and the subjects were followed up for 10 years. The data set contains 4128 subjects. For illustrative purposes, we only chose 1437 female non-Hispanic white patients into our analysis data set. The outcome of interest is time to death, which may be censored due to the disenrollment or study termination.

1.3.4.2 Missing Data Problem Eight covariates are considered in the analysis: age, education level, smoking, duration of diabetes, hemoglobin A1c, major depression, estimated glomerular filtration rate (eGFR), and microalbuminuria. The first six covariates are completely observed, whereas the eGFR and microalbuminuria have some missing values. Among the 1437 selected patients, there are 94 patients with missing eGFR and 400 with missing microalbuminuria. In total, there are 437 patients with missing eGFR or microalbuminuria.

1.3.5 Randomized Trial on Vitamin A Supplement

1.3.5.1 Background Vitamin A is a crucial nutrient required for maintaining immune function, eye health, growth, and survival in human beings. Lack of vitamin A can lead to death. Sommer and Zeger (1997) reported a randomized clinical trial on the effectiveness of a vitamin A supplement in reducing mortality in children in rural Indonesia. In this trial, villages were randomized to either the treatment or the control group. All children in a village assigned to the treatment group would receive vitamin A supplements, and all children in a control village did not receive any vitamin A supplements. Due to various reasons, however, not all children in the treatment villages received a vitamin A supplement. The outcome of interest is the mortality rate within 12 months of the start of the study; that is, whether the children survived at month 12 after the start of the study. In this example, we ignore any clustering within villages.

1.3.5.2 Missing Data Problem When estimating the causal effect of vitamin A supplement, we do not consider which treatment patients would have taken and what outcomes the patients would have had if they had been given treatments different from the ones they have been assigned to. We only observe patient's outcome and actual treatment received under the treatment assignment the patient actually receives.

1.3.6 Randomized Trial on Effectiveness of Flu Shot

1.3.6.1 Background McDonald et al. (1992) studied influenza vaccine efficacy in reducing morbidity in high-risk adults, using a computer-generated reminder for flu shots. The study was conducted over a 3-year period (1978–1980) in an academic primary care practice affiliated with a large urban public teaching hospital. Physicians in the practice were randomly assigned to either an intervention or a control group at the beginning of the study. Since physicians at the clinic each care for a fixed group of patients, their patients were similarly classified. During the study period, physicians in the intervention group received a computer-generated reminder when a patient with a scheduled appointment was eligible for a flu shot. Since a patient whose physician got a computer reminder for flu shot did not need to get a flu shot, some patients in the intervention group did not receive a flu shot. Similarly, some patients in the control group got their flu shots, even though their doctors did not receive a reminder for a flu shot. Hence, noncompliance exists in this data set. This data set has one additional problem of missing data for the outcome variable.

1.3.6.2 Missing Data Problem The reminder information is available on everyone, and we know whether the patients had flu shot. However, we do not know what will happen if the patient had been given a treatment assignment different from the actual treatment assignment the patient received. In addition, flu-related hospitalizations of some patients were not observed.

CHAPTER 2

OVERVIEW OF METHODS FOR DEALING WITH MISSING DATA

Regardless of the existence of missing data, the end result of any analysis is to make valid and efficient inferences about the population of interest. Neyman and Pearson (1933) established valid criteria for evaluating any statistical procedure. These criteria include having a small bias, where bias refers to the difference between the average sample estimate and its true value, and a small variance associated with the average sample estimate (efficiency). Bias and variance can be combined in a single measure called the mean square error (MSE) so that the bias, variance, and MSE describe the behavior of the estimate. Using these criteria, we discuss the various missing data methods that are available, each with its own strengths and limitations. This chapter provides a brief overview of the existing approaches and classifies them according to whether they remove observations with missing data, utilize all available data, or impute missing data. An excellent overview of missing data methods is provided by Schafer and Graham (2002).

2.1 METHODS THAT REMOVE OBSERVATIONS

Many approaches simplify the missing data problem by discarding data. We discuss how most of these approaches may lead to biased results, although some approaches try to directly address this issue. Discarding data also reduces the sample size, which in turn reduces efficiency and leads to larger standard errors for the parameters of interest.

Applied Missing Data Analysis in Health Sciences, First Edition. Xiao-Hua Zhou, Chuan Zhou, Danping Liu, and Xiaobo Ding.

2.1.1 Complete-Case Methods

As the name suggests, the complete-case method (also known as list-wise deletion) makes use of cases with complete data on all variables of interest in the analysis. Cases with incomplete data on one or more variables are discarded. This approach is easiest to implement since standard full data analysis can be applied without modification, and is the default in many statistical software packages (i.e., cases with any missing values are automatically ignored). The disadvantages include loss of precision due to a smaller sample size and bias in complete-data summaries when the cases with missing data differ systematically from the completely observed cases (i.e., when data are not missing completely at random).

In practice, when the loss of precision and the bias are small, we may opt for this approach for its simplicity. Unfortunately, the loss of precision and the bias are difficult to quantify since both depend not only on the fraction of complete cases but also on the extent to which complete and incomplete cases differ on the parameters of interest. So there is no universal rule of thumb, such as a certain percentage of missing data, for deciding whether this is an appropriate method.

2.1.2 Weighted Complete-Case Methods

The earlier complete-case strategy can potentially lead to biased summaries because the sample of observed cases may not be representative of the full sample. Another strategy reweights the sample to make it more representative. For example, if the response rate is twice as likely among men as women, data from each man in the sample could receive a weight of 2 in order to make the data more representative. This strategy is commonly used in sample surveys, especially in the case of unit nonresponse, where all the survey items are missing for cases in the sample that did not participate. For example, if only one variable is missing, this method first creates a model to predict nonresponse in one variable as a function of the rest of the variables in the data. The inverse of the predicted probabilities of response from this model could then be used as survey weights to make the complete-case sample more representative. If more than one variable is missing, this method becomes more complicated, and there will be problems with standard errors if predicted response probabilities are close to 0.

We give an example to illustrate this method more formally. Let Y be the outcome of interest and R be the indicator of whether Y was observed. We also let X denote the set of variables that were observed for both respondents and nonrespondents. Suppose that the MAR (missing at random) assumption holds, that is, $\Pr(R|X, Y) = \Pr(R|X)$. Then the response probability of the ith subject is denoted by

$$\pi(X_i) = \Pr(R_i = 1|X_i).$$

If $\pi(X_i)$ values are known for all n subjects in the sample, we could use

$$\sum_{i=1}^{n} \frac{R_i}{\pi(X_i)} Y_i$$

to estimate the marginal mean of Y, which is the so-called inverse probability weighted estimator (Cassel et al., 1983; Skinner and D'arrigo, 2011). If the parameter θ is interest, which satisfies $E[g(Y, X; \theta)] = 0$, we could solve the following estimating equation to find an estimate of θ:

$$\frac{1}{n} \sum_{i=1}^{n} \frac{R_i}{\pi(X_i)} g(Y_i, X_i; \theta) = 0.$$

In many practical applications, the response probability $\pi(X)$ is unknown. In this case, we may assume a parametric model for it, that is, $\pi(X)$ is known up to some parameters. Or we can estimate $\pi(X)$ nonparametrically if X does not contain many variables. In some applications, we may group the values of $\pi(X)$ into four to six categories and create a new variable C, which is used to adjust weights in the analysis. This might be required because directly weighting by $[\pi(X)]^{-1}$ places more reliance on correct model specification of the regression of R on X than adjusting for C (Little and Rubin, 1987).

2.1.3 Removing Variables with Large Amounts of Missing Values

This approach drops variables from the analysis that have a large proportion of missing values. This is not recommended since it may exclude important variables in the regression model, required for causal interpretation, and may lead to bias and unnecessarily large standard errors. For important variables with more than trivial amounts of missing data, the explicit model-based methods in the next sections are suggested.

2.2 METHODS THAT UTILIZE ALL AVAILABLE DATA

In contrast to complete-case methods, available case methods (also known as pairwise deletion methods) utilize information from both completely and partially observed cases. In multivariate regression, cases with incomplete data on certain variables can still provide information about the relationship between the outcome and other observed variables. For example, in repeated measures data, cases with incomplete data at one time point can still provide valid information about the relationship between variables at other time points. Mixed effects models are therefore a popular choice for dealing with longitudinal data with missing outcomes. Because the available case methods make use of more information than the complete-case methods, they are generally better at correcting for bias when data are MAR than the complete-case methods.

A major disadvantage of this approach is that different analyses would use different subsets of the data, depending on the variables used and their missing data patterns, which may not provide consistent inference. And similar to complete-data methods, there will be a loss of efficiency from discarding the partially observed variables, and the available case summaries will be biased when the cases with missing data differ systematically from the observed cases.

2.2.1 Maximum Likelihood

Many procedures arise from defining a model for the variables with missing values and making statistical inferences based on *maximum likelihood* (ML) methods. The fundamental idea behind the ML methods is conveyed by their name: find the values of the parameters that are most probable, or most likely, for the data that have actually been observed.

Denote D_{obs} as the observed data, which have the probability density $p(D_{\text{obs}}|\theta)$. Here θ is a set of parameters of interest. Then, the *likelihood function* $L(\theta|D_{\text{obs}})$ is equal to $p(D_{\text{obs}}|\theta)$, that is, $L(\theta|x) = p(D_{\text{obs}}|\theta)$, and is thought of as a function of θ, where the data D_{obs} are fixed. The *log-likelihood* $l(\theta|D_{\text{obs}})$ is equal to $\log(L(\theta|x)$ and is, in general, easier to be maximized.

The next step is to deduce a numerical value for θ using our knowledge of $L(\theta|x)$. The ML principle says we should choose an estimate of θ that maximizes the log-likelihood function. In other words, we choose a value of the parameter that best explains the observed data, and this value is called the *ML estimate* of θ.

Finding the maximum can be easy or hard, depending on the form of $p(D_{\text{obs}}|\theta)$. For the normal (Gaussian) distribution, θ consists of the mean μ and variance σ^2, and we can find the derivative of $l(\theta|D_{\text{obs}})$, set it equal to zero, and solve directly for μ and σ^2. But for most problems, these analytic expressions are hard to come by, and we will have to use more elaborate techniques. One might use the Newton–Raphson or quasi-Newton algorithms to maximize directly the likelihood of the observed data. One of the most popular techniques for maximizing the likelihood function is the expectation–maximization (EM) algorithm of Dempster et al. (1977). We briefly discuss the EM algorithm in Section 2.2.1.1 and readers may find the theoretical details in Little and Rubin (1987).

2.2.1.1 EM Algorithm The EM algorithm is a general method of finding the ML estimate of the parameters of an underlying distribution from a given data set when the data are incomplete or have missing values.

The EM algorithm has two main applications: First, to find the ML estimates when data truly have missing values (e.g., patients lost to follow-up); Second, to find the ML estimates when the likelihood is analytically intractable and difficult to be optimized, but the likelihood can be simplified by assuming that there exist additional, but missing parameters. The algorithm can be applied whether the missing data are ignorable or not by including a missing data model in the likelihood.

The EM algorithm is essentially an iterative optimization algorithm that will, under certain conditions, converge to parameter values at a local maximum of the likelihood function. Let D_{mis} be the missing data. Then, the full data $D_{\text{com}} = \{D_{\text{obs}}, D_{\text{mis}}\}$. In general, the full data log-likelihood will have an easily defined, analytically solvable maximum, but maximization of the observed-data log-likelihood $l(\theta|Y_{\text{obs}})$ may have no analytic solution.

The EM algorithm approach is a procedure that iterates between two steps: an expectation step (E step) and a maximization step (M step). The E step calculates the expected value of the full data log-likelihood, given the observed data and current estimated parameters. The M step performs maximization on the expectation of the full data log-likelihood, computed in the E step above. The steps are usually easy to construct programming, and implementation of the EM algorithm is straightforward using standard statistical software packages.

2.3 METHODS THAT IMPUTE MISSING VALUES

In contrast to methods that remove observations with missing values, imputation methods impute or ''fill in'' missing values. Perhaps because of long-standing taboos against ''making up'' data, it has taken some time for imputation methods to gain general favor among the research community despite being as sound as other acceptable approaches. The key realization behind imputation is that, although the missing values are not observed, information about these values may still be extracted from the other observed variables.

This section will consider methods that impute, or fill in, missing values. These methods can be implemented by imputing one value for each missing item (single imputation), or by imputing more than one value to allow for appropriate assessment of imputation uncertainty (multiple imputation). In general, anytime a single imputation strategy is used, standard errors of estimates of parameters will be underestimated since the resulting data are treated as a complete sample, ignoring the consequences of imputation. In contrast, multiple imputation strategies take into account the uncertainty about the imputed missing values, and are therefore preferable to single imputation. The single and multiple imputation strategies are discussed in Sections 2.3.1 and 2.3.2, respectively.

2.3.1 Single Imputation Methods

2.3.1.1 Unconditional Mean Imputation A popular approach replaces each missing value with the unconditional mean of the observed values for a particular variable. This is a single imputation approach and can therefore lead to underestimating standard errors. Additionally, this method distorts relationships between variables since it pushes the correlations between variables toward zero (Gelman and Hill, 2006).

2.3.1.2 Conditional Mean Imputation For data sets with few missing variables, conditional mean imputation (also known as regression imputation) is an improvement on the unconditional mean approach since it replaces each missing value with the conditional mean of the variable based on other fully observed variables in the data set. If there is only one variable with missing values, say Y, then a regression model for predicting Y from the other variables $X = (X_1, \ldots, X_k)'$ can be used to impute Y. The first step fits the model to the cases where Y is observed. The second step plugs in the X values for the nonrespondents into the regression equation, obtaining predicted values $\hat{Y}$ for the missing values of Y. This is a single imputation approach and can therefore lead to underestimating standard errors, although this approach is acceptable for some problems if the standard errors are corrected (Schafer and Schenker, 2000). However, it is not recommended for analyses of covariances and correlations since it distorts the strength of the relationship between X and Y. This approach is especially problematic when one is interested in the details of the distribution (Little and Rubin, 1987). For example, an approach that imputes missing incomes using conditional means tends to underestimate the percentage of cases in poverty.

2.3.1.3 Last Observation Carried Forward (LOCF) For longitudinal studies, this approach imputes the missing value at one time point with its value at the previous time point, or uses the last observation carried forward. Both mean imputation and LOCF are not recommended since they can potentially induce bias as well as underestimate variability (Sasha and Jones, 2009).

2.3.1.4 Substitution of Related Observations Other methods may utilize data from related observations. For example, suppose that we are missing the education level of some of the fathers of children. We might fill in these values using the mothers' report of the fathers' education level. This may seem plausible, but it introduces measurement error, especially if there is a reason to believe that the reporting subjects may misrepresent the measurements of the people about whom they are providing information. As another example, if a household cannot be contacted, we might substitute their missing values with values from a household in the same neighborhood.

2.3.1.5 Hot Deck Imputation This approach selects a value at random from the pool of "similar" complete cases, and the missing value is replaced by the selected value. An advantage of this approach is that it does not require careful modeling to develop the selection criteria for imputing the value, although bias can still be introduced. Note that this approach can serve as a multiple imputation method by selecting multiple values at random from the pool.

2.3.1.6 Missing Indicator Method Some *ad hoc* approaches seem to solve the missing data problem by retaining the data, but in reality, merely redefine the parameters of the population. Consider a linear regression of Y on X, where X has some missing values. Suppose we recode all missing values of X to a common number and create a binary indicator R, which is equal to 1 when X is missing and 0 otherwise. In our original model, $E(Y|X) = \beta_0 + \beta_1 X$, where β_0 and β_1 represent the intercept and the slope of the full population, respectively. Adjusting for missing indicator R, the model becomes $E(Y|X) = \beta_0 + \beta_1 X + \beta_2 R$, where β_0 and β_1, respectively, become the intercept and slope among respondents and $\beta_0 + \beta_2$ becomes the intercept for nonrespondents. In particular, this method forces the slope to be the same across groups with and without missing X, which will likely lead to biased coefficient estimates.

Adding an interaction in the model can help lessen this bias, that is, replace the term $\beta_1 X$ with $\beta_1 X(1 - R)$, and this approach leads to estimates similar to the complete-case estimates. But in general, these approaches are not recommended because of the potential bias they induce in their attempt to estimate population-level summaries (Jones, 1996; Greenland and Finkle, 1995).

For categorical variables with missing values, this approach creates an additional category for missing values. In the regression example above, if X is a nominal variable with values $1, \ldots, k$, then treating the missing value as category $k + 1$ merely redefines categories $1, \ldots, k$ to apply to respondents only. This approach can lead to biased results, even if the data are missing completely at random (MCAR).

2.3.2 Multiple Imputation

While imputation may solve the missing data problem, naive or unprincipled imputation methods may create more problems than they solve, by distorting estimates and standard errors. Rubin (1987) addressed the question of how to obtain valid inference from imputed data by using multiple imputation, a technique where missing data are replaced by m values drawn from the posterior predictive distribution of the missing data, resulting in m complete data sets. Standard full data methods are then used on each of the m complete data sets to estimate the parameters of interest, resulting in m estimated parameters and their corresponding standard errors. These estimates are then combined, or averaged, to produce estimates and confidence intervals that incorporate the missing data uncertainty. This averaging will, in general, lower the variance of the combined estimate. In particular, this method recognizes two sources of uncertainty: the uncertainty caused by sampling the subjects from a source population, and the uncertainty in estimating underlying distributions of the variables with missing values. A simple method is used to combine both sources of uncertainty, resulting in a single corrected standard error of the estimated association.

The multiple imputation method was originally developed in the setting of large public-use data sets from sample surveys and censuses. However, with advances in computational methods and software for creating multiple imputations, the technique has become attractive to researchers in most fields where missing data cause a problem, as documented in Schafer (1997b).

Software most readily available to implement multiple imputation assumes that the missing data are MAR, although the paradigm certainly does *not* require MAR nonresponse. In principle, multiple imputations can be created in any kind of model for the missing data mechanism, and resulting data will be valid under that mechanism (Rubin, 1987). But methods that assume MAR have the advantage of avoiding explicit modeling of nonresponse, although unfortunately the observed data cannot provide enough information on the validity of the MAR assumption.

Usually the imputed values are drawn from an imputation model. If only one variable is subject to missingness, it is easy to construct the imputation model. For example, we can use logistic regression to model a binary variable, and a linear

Gaussian model for a continuous variable. If there are multiple variables subject to missingness, the imputation model can be constructed jointly or conditionally.

The joint modeling approach assumes that given the other observed variables, the variables subject to missingness follow a multivariate normal distribution. However, in practice, the variables subject to missingness may involve different types such as binary, categorical, and ordinal variables. In this case, it would be computationally implausible to simulate the posterior predictive distribution simultaneously for all the missing variables. The multiple imputation with chained equations is proposed. The idea of chained equations is to update the missing variables one by one, based on a series of full conditional distributions, as proposed by van Buuren and Groothuis-Oudshoorn (2011).

Denote $Y = (Y_1, \dots, Y_p)'$ the set of p variables in the data set that are subject to missingness, which include both response variables and covariates. Let $Y_{-j} = (Y_1, \dots, Y_{j-1}, Y_{j+1}, \dots, Y_p)'$ be the collection of all variables in Y but Y_j. The model assumptions are made for each variable Y_j fully conditional on all other variables Y_{-j}:

$$
\begin{gathered}
P(Y_1 \mid Y_{-1}, \theta_1) \\
\vdots \\
P(Y_p \mid Y_{-p}, \theta_p).
\end{gathered}
$$

This model framework is called "chained equations". It specifies p univariate regression models, and avoids the hassle of jointly model the multivariate distributions for all the missing variables.

With a sample of size n, the observed data on Y_j is a vector of observations from n subjects, denoted by $\boldsymbol{y}_j$. Hence we can partition $\boldsymbol{y}_j$ into the missing part $\boldsymbol{y}_j^{\text{mis}}$ and the observed part $\boldsymbol{y}_j^{\text{obs}}$, indicating a pool of subjects with observed or missing Y_j. It should be noted that this is a partition of subjects for each variable, whereas the partition of variables for a subject will be introduced in later chapters. The iteration procedures are described as follows:

$$
\begin{aligned}
\theta_1^{*(t+1)} &\sim P(\theta_1 \mid \boldsymbol{y}_1^{\text{obs}}, \dots, \boldsymbol{y}_p^{(t)}), \\
\boldsymbol{y}_1^{*(t+1)} &\sim P(\boldsymbol{y}_1^{\text{mis}} \mid \boldsymbol{y}_1^{\text{obs}}, \boldsymbol{y}_2^{(t)} \dots, \boldsymbol{y}_p^{(t)}, \theta_1^{*(t+1)}), \\
&\vdots \\
\theta_p^{*(t+1)} &\sim P(\theta_p \mid \boldsymbol{y}_p^{\text{obs}}, \boldsymbol{y}_1^{(t+1)}, \dots, \boldsymbol{y}_{p-1}^{(t+1)}), \\
\boldsymbol{y}_p^{*(t+1)} &\sim P(\boldsymbol{y}_p^{\text{mis}} \mid \boldsymbol{y}_p^{\text{obs}}, \boldsymbol{y}_1^{(t+1)} \dots, \boldsymbol{y}_{p-1}^{(t+1)}, \theta_p^{*(t+1)}),
\end{aligned}
$$

where $\boldsymbol{y}_j^{(t)} = (\boldsymbol{y}_j^{\text{obs}}, \boldsymbol{y}_j^{*(t)})$ is the imputed data vector of $\boldsymbol{y}_j$ after the tth iteration. The draws for the parameters are just a full Bayesian regression analysis. Note that for drawing $\theta_j^{*(t+1)}$, the condition is on $\boldsymbol{y}_j^{\text{obs}}$ rather than $\boldsymbol{y}_j^{(t)}$. In other words, only subjects with observed response (Y_j) are included in the regression; the covariates (Y_{-j}) are all imputed based on the previous iteration. The draws for the missing data are performed sequentially for each variable on the subjects with missing values, which is based on the p univariate regression models. This procedure has been implemented in the `R` package `mice`.

In practice, the chained equation approach can be briefly broken down into the following steps, and we will illustrate these with examples in the later chapters.

1. Examine the missing data patterns and identify key variables with missing values that require imputation. Decide on the sequence of the imputation, usually based on amount of missingness and quality of the variables used to impute them.
2. Initialize all missing values with a simple imputation, such as mean imputation or last-value-carry-forward.
3. Construct an imputation model (typically a regression model) for the first variable (Y_1) requiring imputation. For example, if Y_1 is a binary variable, a logistic regression model may be constructed with Y_1 being the dependent variable, and other variables as covariates. For other types of variables, linear, Poisson or other generalized linear regression models may be employed.
4. The missing values for Y_1 are then replaced with predictions (imputations), using either predictive mean matching method or direct sampling from predictive posterior distributions. Note the latter method may require Gibbs sampling for intractable distributions. When Y_1 is used in subsequent imputation models for other variables, the observed values and imputed values for Y_1 will be combined to form a complete data vector.
5. Steps 3 and 4 are repeated for each variable that has missing values, $Y_2, \ldots, Y_p$. The cycling through each of the variables constitutes one iteration. At the end of each iteration all of the missing values have been replaced with imputed values from regressions that are assumed to reflect relationships among these variables. The number of iterations to be performed can be decided by the researchers. Though it is suggested that 10 iterations are usually sufficient, nowadays with the increasing computing power, we recommend performing iterations many more times to ensure convergence of the algorithm. The imputed data set in the last iteration is taken to be one imputed data set.
6. Iterate steps 2 through 5 to obtain multiple imputed datasets. The number of imputations should depend on the missing information of the data set. In most cases, 10-20 imputations would be sufficient.
7. With each imputed complete dataset, we then proceed with parameter estimation. The results are pooled together over all imputed datasets for the final inference.

The chained equation is able to handle flexible type of missing variables, and is usually simple to implement. Although the convergence of the algorithm is not proved rigorously, its performance in practice is stable in most circumstances, as long as the missing data proportion is not too high. One important limitation of the chained equation approach is that there is no guarantee that the imputed sample is indeed drawn from the posterior predicted distribution because there may not exist a joint distribution that yields the specified conditional distributions.

2.4 BAYESIAN METHODS

Bayesian statistical conclusions about the parameter θ or missing data D_{mis} are made in terms of probability statements that are conditional on the observed value D_{obs}, or $p(\theta \mid D_{\mathrm{obs}})$ and $p(D_{\mathrm{mis}} \mid D_{\mathrm{obs}})$. It is exactly this conditioning on the observed data that makes Bayesian inference different from the commonly used statistical inference described in most articles and textbooks, which is based on an evaluation of the procedure used to estimate θ or D_{mis} over the distribution of possible D_{obs} values conditional on the true unknown value of θ. Note that in simple settings, results from both methods can be quite similar.

Bayesian methods make no distinction between the missing data and the parameters since both are uncertain. Focus is placed on the joint distribution of the missing data and parameters conditional on observed data, and the typical setup for the joint distribution includes a prior distribution for the parameters, a joint model for all the data (observed and missing), and a model for the missing data process (unless the missing data mechanism is ignorable, in which case this last model is not necessary).

Missing variables are sampled from their conditional distribution via different methods such as the Gibbs sampler. There is an imputation stage where simulated draws from the posterior predictive distribution of unobserved values D_{mis} conditional on observed values D_{obs} are imputed. Then there is a draw from the posterior distribution of the parameter θ. Inference then proceeds by averaging over the distribution of missing values. Fully Bayesian methods with missing variables only involve incorporating an extra step in the Gibbs sampler, compared with the case of having no missing values in the data. Therefore, Bayesian methods do not require additional techniques for inference to accommodate missing data. For this reason, Bayesian methods are quite powerful.

CHAPTER 3

DESIGN CONSIDERATIONS IN THE PRESENCE OF MISSING DATA

It should be pointed out that researchers themselves should be blamed in part for the high prevalence of missing data in biomedical, behavioral, and social research. Many missing data problems arise from a poor study design and lack of careful planning. They can be largely avoided if enough attention is given at the beginning of the study. As Benjamin Franklin put it, "an ounce of prevention is worth a pound of cure." So a small amount of planning ahead can often lead to greatly reduced bias and improved efficiency, sometimes can even salvage an otherwise wasted effort. In this chapter, we outline some design and conduct strategies to avoid or reduce missing data in biomedical research studies. Most of the advice is based on a recent National Research Council report (National Research Council, 2010) and a special report in the *New England Journal of Medicine* (Little et al., 2012b). More technical information can be found in Little et al. (2012a).

3.1 DESIGN FACTORS RELATED TO MISSING DATA

The best advice regarding missing data is probably the one given by R.A. Fisher, "the best solution to handle missing data is to have none." This may sound intuitive; however, it is very difficult to achieve in practical settings. In fact, for most clinical studies, more often than not there will be missing data. Some researchers might put little effort in study designs and data collection to prevent missing data, since the

Applied Missing Data Analysis in Health Sciences, First Edition. Xiao-Hua Zhou, Chuan Zhou, Danping Liu, and Xiaobo Ding.

missing data problem is so pervasive that it is deemed unavoidable. However, we believe that every effort should be made early to reduce missing data. Not only does this retain the most information for the estimation and inference, it also encourages the researchers to think more carefully about their scientific question, study design, data collection, and analysis.

To avoid missing data, it is helpful to understand what factors could affect the availability of data. In the setting of clinical trials, the following factors tend to be highly influential:

1. *The Ease of Assessment and Collection of the Variables*: The occurrence of missing values is expected to be lower when the outcome variable is easy to assess, such as mortality (e.g., cardiovascular trials) than when the outcome is more difficult to assess and requires the active participation of patients and/or sophisticated methods of diagnosis.
2. *The Quality of the Definition of the Variables*: Subjects respond better to clearly defined outcome measures than those vaguely defined.
3. *The Nature of the Variables*: Subjects respond better with respect to less personal or sensitive variables. Variables such as HIV infections, psychiatric disorder, and income are hard to collect.
4. *The Amount of Data Collected on Individuals*: Missing data become more frequent when the respondents feel that the data collection process is too burdensome.
5. *The Duration of the Clinical Trials*: The longer the follow-up, the greater the likelihood of missing values.
6. *The Therapeutic Indication*: Missing values are more frequent in those diseases where the adherence of patients to the study protocol is usually low (e.g., psychiatric disorders).
7. *The Treatment Modalities*: For example, surgical versus medical treatment.
8. *Motivation of Participants to Remain in the Study*: Patients who are more tolerable to the intervention or more likely to benefit from the study tend to remain within the study.

When planning a study, the researchers need to evaluate the above-mentioned factors and try to come up with ways to ease their impact on creating missing data. Several major difficulties arise as a result of the presence of missing values, and these are aggravated as the number of missing values increases. Thus, it is extremely important to avoid the presence of unobserved measurements as much as possible. Researchers should favor designs that minimize this problem, as well as strengthen data collection regardless of the patient's adherence to the protocol and encourage the retrieval of data, such as continued access to medical records after the patient's dropout. In some circumstances, in particular where retrieved dropout information represents the progression of the patient without (or before) impact of further therapeutic intervention, these data will give the best approximation to the intention to treat (ITT) population.

3.2 STRATEGIES FOR LIMITING MISSING DATA IN THE DESIGN OF CLINICAL TRIALS

As pointed out by Lavori et al. (2008), a good study design may not completely eliminate the problem of missing data; however, it can reduce the amount of missing data or their impact. Subsequently, statistical methods such as those covered in this book can be applied at the analysis stage to make proper estimation and inference. On the other hand, if insufficient attention is paid to missing data at the design stage, then the damage to the data may be so severe that the inferential problems are impossible to be resolved in the statistical analysis phase.

There are design elements for clinical trials that can help to prevent missing data by increasing patient participation, reducing patient dropouts, and improving data collection. We recommend the following design strategies to reduce missing data in clinical trials; many of them have been discussed in National Research Council (2010). Note that much of the advice also applies to observational studies.

1. **Improve Data Collection**
   - Focus on the research objectives and only collect data that are absolutely necessary to address the scientific questions. Researchers are often tempted to collect as much information as possible. But too much unnecessary data collection tends to put stress on resources, staff, and participants.
   - Design the data collection carefully to focus on the most pertinent information so that subjects will not be overwhelmed with reporting.
   - Provide clear definitions to each of the variables to avoid ambiguity. Missing data often occur when the respondents are unclear about a question or variable.
   - Consider multiple modes of data collection. In addition to participants reporting, administrative data set, electronic medical records, census data, and online surveys can also be used.
   - For longitudinal or repeated measures studies, give extra flexibility to the data collection schedule.
   - Plan the study period carefully so that it is not unnecessarily long, which can lead to more dropouts. For example, shorten the follow-up period for the primary outcome.
   - Identify variables that are most likely to be missing, and plan to spend extra effort to collect them.
2. **Improve Study Population Selection**
   - Select the study population carefully. Target the population that is not adequately served by current treatments and hence has an incentive to remain in the study. Subjects with more severe disease or poor health status are more likely to drop out or not respond. However, be mindful that such an exclusion may limit generalizability of the study findings.

- Include a run-in period during which all patients are assigned to the active intervention, and after which only those who tolerated and adhered to the therapy undergo randomization. The potential drawbacks are carry-over effects and unblinding of patients to intervention.

3. **Improve Intervention**
   - Consider add-on interventions, in which a study treatment is added to an existing treatment, typically with a different mechanism of action known to be effective in the previous studies.
   - Allow the rescue medications that are designated as components of a treatment regimen in the study protocol.
   - Provide continued access to effective treatments after the trial, before treatment approval.

3.3 STRATEGIES FOR LIMITING MISSING DATA IN THE CONDUCT OF CLINICAL TRIALS

The incidence of missing data varies greatly across studies. Some of this variation is context specific, but in many cases more attention to planning and conduct can substantially reduce the problem. Some of the strategies are listed in the following. Many of the ideas are discussed in more detail in the panel report of Little et al. (2012a,b).

1. **Better Training of Investigators, Staff, and Patients**
   - Train investigators and research staff so that they can keep participants in the trial until the end, regardless of whether the participants continue to receive the assigned treatment.
   - Select investigators and clinical research associates with good track record of enrolling and following participants and collecting complete data in the previous trials.
   - Patients also need to be educated about the importance of providing complete data even if they choose not to follow the assigned treatment.
2. **Better Data Tracking**
   - Set acceptable target rates of missingness in the data and monitor the progress of the trial with respect to these targets.
   - Such targets and monitoring can also be tied to the research staff's performance evaluation.
   - Apply double or multiple data entry or other data checking methods to identify data issues on an ongoing basis. Identifying missing data early may allow the research staff to recollect data in time.
   - Keep contact information for participants up to date. Collect multiple contact numbers or e-mails during enrollment so that participants are more likely to be located if they move out of the study area.

- Collect information from participants regarding the likelihood that they will drop out, and use this information to attempt to reduce the incidence of dropout. For example, the insurance status or the employment status tends to have a strong influence on participant's decision to remain in the study.

3. **Better Patient Experience**
    - Limit the burden and inconvenience of data collection on the participants. For example, allowing participants to fill out Web-based surveys at their own time and pace may be preferred than paper surveys.
    - Incentives often help, but there could be political, financial, or administrative restrictions.
    - Make the study experience as positive as possible. Sometimes small things such as free parking or friendly greeting from the research staff can make big differences.

3.4 MINIMIZE THE IMPACT OF MISSING DATA

There is no rule regarding the maximum number of missing values that could be acceptable. Although the conventional wisdom is that more than 30% of missingness should raise an alarm to the analyst, the impact of missing data is determined by the missing information rate, which depends on both the sample size and the statistical model for complete data. See Little and Rubin (2002) for a detailed discussion on the definition of the missing information rate.

It is very important when designing the study and specifying the statistical methods to be used, to anticipate the number of missing values likely to be observed in the trial. Experience from exploratory trials and from trials of other products in similar indications should inform expectations for missing data when planning the trial. Careful planning will help specify both a plausible approach to handling missing data and a range of sensitivity analyses that could explore the impact of departures from the expected missing data pattern. Indeed, an estimate of the foreseen and acceptable amount of missing data is highly recommended for the following reasons: first, this may have repercussions for the variability and the expectations of the effect size and hence the sample size calculation; second, proper planning will minimize the risk that the strategy for missing data handling itself introduces bias; and third, the uncertainty in interpreting the results introduced increases (and hence the number of sensitivity analyses required may need to increase as the number of missing values increases).

There is no universally applicable method of handling missing values, and different approaches may lead to different results. To avoid concerns over data-driven selection methods, it is essential to prespecify the selected methods in the statistical section of the study protocol or analysis plan. This section must include a detailed description of the selected methods and a justification on why the methods to be applied are expected to be an appropriate way of summarizing the efficacy results of the study and to result in the absence of bias in favor of the experimental treatment. The sensitivity analyses to be performed should also be prespecified.

When missing data are unavoidable, there are ways to reduce their impact on the inference:

- Thoroughly consider all possible factors that may result in missing values.
- Collect auxiliary data that could help imputation of missing values.
- Collect secondary outcomes that capture similar information as primary outcomes, but are easier to collect.
- Where possible, outcome data after withdrawal should be collected. Also, data should be collected on other therapies received post dropout. Specifically, full details of the type of therapy given, including when and for how long it was used and at what dose, should be collected. This information will allow the value of any outcome data collected after withdrawal to be put into context.

CHAPTER 4

CROSS-SECTIONAL DATA METHODS

4.1 OVERVIEW OF GENERAL METHODS

As the name suggests, cross-sectional data analysis refers to the analysis that looks at the data collected on subjects at one time point, or without regard to time differences (a single-occasion snapshot of the system of variables and constructs), and analyses typically consist of comparing differences between subjects. Even with longitudinal data, cross-sectional analysis is often performed by looking at individual time points. For example, in the IMPACT study described later, investigators randomized patients into two groups and examined depression outcomes at 6 months after baseline to determine if average depression scores differed between the two groups. Many of the missing data methods developed for this setting have been discussed in Chapter 2. Among the recommended approaches that are further described in this chapter, there are four main themes: maximum likelihood approach, Bayesian methods, multiple imputation (MI), and inverse probability weighting (IPW). In addition, we discuss some more advanced techniques such as doubly robust estimators toward the end of the chapter.

4.2 DATA EXAMPLES

We will illustrate the methods using simulated data as well as data from real-world studies. Three real-world data sets are considered: the NHANES study, the IMPACT

Applied Missing Data Analysis in Health Sciences, First Edition. Xiao-Hua Zhou, Chuan Zhou, Danping Liu, and Xiaobo Ding.

study, and the NACC study. The last two examples have already been described in Chapter 1. Next, we briefly describe the simulation study and the NHANES study.

4.2.1 Simulation Study

To illustrate the missing data problems and compare various approaches, we will use simulated data examples with known missing data mechanisms imposed. Scenarios considered will include mixed variable types, missing outcomes only, missing covariates only, and missing both outcomes and covariates.

4.2.2 NHANES Study

We will use a small data set from NHANES with nonmonotone missing values, which has been included as a working example with the R `mice` package (van Buuren and Groothuis-Oudshoorn, 2011). The data set was first used by Schafer (1997b), and contains 25 observations on the following four variables: age (age group, $1 = 20\text{–}39, 2 = 40\text{–}59, 3 = 60+$), bmi (body mass index, km/m^2), hyp (hypertensive, $1 =$ no, $2 =$ yes), and chl (total serum cholesterol, mg/dl). The variables age and hyp are considered discrete variables, whereas bmi and chl are treated as continuous variables. We will consider linear models and generalized linear models with bmi and hyp as outcomes.

4.3 MAXIMUM LIKELIHOOD APPROACH

Let D_{com} denote the complete-data values in the absence of missing values. We write D_{obs} and D_{mis} as the observed and missing parts of D_{com}, respectively. We then base our inference on the observed-data likelihood function under the ignorable missing data mechanism. The observed likelihood function can be calculated as follows:

$$L(\theta \mid D_{\text{obs}}) = \int f(D_{\text{obs}}, D_{\text{mis}} \mid \theta) \mathrm{d}D_{\text{mis}}.$$

Its score equation can also be obtained as follows:

$$D_1(\theta \mid D_{\text{obs}}) = \frac{\partial \log L(\theta \mid D_{\text{obs}})}{\partial \theta} = 0.$$

The maximum likelihood estimator (MLE) for θ can be found by solving the score equation.

In general, there is no closed-form solution for the score equation, and we need to use an iterative method for obtaining a solution for θ. Two commonly used iterative methods are the Newton–Raphson method and the Fischer scoring method. Let $\theta^{(t)}$ be the current value of θ after t iterations. The Newton–Raphson method defines the next iteration value $\theta^{(t+1)}$ as follows:

$$\theta^{(t+1)} = \theta^{(t)} + I^{-1}(\theta^{(t)} \mid D_{\text{obs}}) D_1(\theta^{(t)} \mid D_{\text{obs}}),$$

where $I(\theta \mid D_{\text{obs}})$ is the observed information matrix, defined as

$$I(\theta \mid D_{\text{obs}}) = -\frac{\partial^2 l(\theta \mid D_{\text{obs}})}{\partial\theta\partial\theta},$$

where $l(\theta \mid D_{\text{obs}}) = \log L(\theta \mid D_{\text{obs}})$. The Fisher scoring method gives the following iterative value:

$$\theta^{(t+1)} = \theta^{(t)} + J^{-1}(\theta^{(t)} \mid D_{\text{obs}}) D_1(\theta^{(t)} \mid D_{\text{obs}}),$$

where $J(\theta \mid D_{\text{obs}})$ is the expected information matrix, defined as

$$\begin{aligned} J(\theta) &= E\{I(\theta \mid D_{\text{obs}}) \mid \theta\} \\ &= -\int \frac{\partial^2 l(\theta \mid D_{\text{obs}})}{\partial\theta\partial\theta} f(D_{\text{obs}} \mid \theta) \mathrm{d}D_{\text{obs}}. \end{aligned}$$

A major limitation of these algorithms is that they require computation of the matrix of second derivatives of the log-likelihood. An alternative iterative method is the expectation–maximization (EM) algorithm, which is a general method for finding the MLE with missing data and does not require computation of second derivatives of the observed-data likelihood (Dempster et al., 1977). The basic idea of the EM algorithm for dealing with missing values is that it replaces missing values with estimated values, estimates parameters based on the pseudo-complete data, and iterates through these two steps until convergence. Formally, the EM algorithm consists of an E (expectation) step and an M (maximization) step. An additional advantage of the EM algorithm is that both the E and M steps can oftentimes be easily computed. We now give a mathematical definition of EM.

We let $l(\theta \mid D_{\text{com}})$ be the complete-data log-likelihood function. Let $\theta^{(t)}$ be the current estimate after t iterations. The $t + 1$ iteration of the EM algorithm is defined as follows.

- *E Step:* Compute the conditional expectation of the complete-data log-likelihood function, given the observed data and the current parameter value as follows:

$$Q(\theta \mid \theta^{(t)}) = E(l(\theta \mid D_{\text{mis}}, D_{\text{obs}}) \mid D_{\text{obs}}, \theta = \theta^{(t)}).$$

- *M Step:* Maximize this conditional expected log-likelihood $Q(\theta \mid \theta^{(t)})$ with respect to θ to obtain the next iterative value $\theta^{(t+1)}$.

Because we can relate the MLE from the observed-data likelihood to the complete-data likelihood, the EM algorithm has nice theoretical properties. One such property is that every iteration of the EM algorithm increases $l(\theta|D_{\text{obs}})$, that is,

$$l(\theta^{(t+1)}|D_{\text{obs}}) \geq l(\theta^{(t)}|D_{\text{obs}}),$$

with equality if and only if

$$Q(\theta^{(t+1)}|\theta^{(t)}) = Q(\theta^{(t)}|\theta^{(t)}).$$

This means that the likelihood function increases at every iteration of the EM algorithm until the condition for equality is satisfied and a fixed point of the iteration is reached. If the likelihood is bounded (which might happen in many cases of interest), the EM algorithm yields a nondecreasing sequence, which must converge as the number of iterations goes to infinity. Under fairly general conditions, the resulting fixed point will be the MLE. For proofs and other details, see McLachlan and Krishnan (1998) and Wu (1983). For example, if the likelihood function is unimodal and satisfies a certain differentiability condition, then any EM sequence will converge to the unique maximum likelihood estimate, irrespective of its starting value (Wu, 1983). But for likelihood functions with multiple maxima, convergence might go to a local maximum depending on the starting value.

Although the EM algorithm has nice theoretical properties, it also has some limitations. For example, its convergence is usually much slower than the Newton-Raphson method, and it does not provide a variance–covariance matrix estimate of the resulting MLE. In addition, sometimes the M or E step can be difficult to implement. With today's ever more powerful computing environment, the M step can be easily carried out using existing software packages with increasingly efficient numerical optimization algorithms. The biggest obstacle preventing the EM algorithm from being more widely used when confronted with missing data problems is the difficulty in calculating the expectation of the complete-data likelihood. The original motivation for the EM algorithm is that the complete-data log-likelihood is more tractable than the observed–data log-likelihood. However, for the E step, it is often difficult to find a closed form of the expectation of the complete-data likelihood for many practical problems. Numerical methods such as Laplace approximation and importance sampling can be easily implemented. For more complex distributions, advanced sampling algorithms such as the adaptive rejection sampling technique can be applied. Another situation when the EM algorithm may not be a good option is that when interactions or transformations of variables with missing values are present in the regression. In order to overcome these limitations, some extensions have been proposed, including ECM, AECM, PX-EM, and ECME. See McLachlan and Krishnan (1998) for a detailed description of these extensions.

4.3.1 EM Algorithm for Linear Regression with a Missing Continuous Covariate

Now let us demonstrate the EM algorithm with a linear regression model when a continuous covariate is missing. Consider the following regression model:

$$Y = \beta_0 + \beta_1 X_1 + \beta_2 X_2 + \varepsilon, \tag{4.1}$$

where $\varepsilon \sim N(0, \sigma^2)$. Here, Y and X_2 are completely observed, whereas X_1 is possibly missing for some subjects. We assume that $X_1 \mid X_2 \sim N(\alpha, \delta^2)$. Note that α and δ could be functions of X_2 based on the conditional model assumption $X_1 \mid X_2$. Suppose that we have a simple random sample of size n from model (4.1). Without loss of generality, we assume that the first m subjects have missing X_1 values, whereas the remaining subjects are completely observed. Write $\boldsymbol{y} = (y_1, \ldots, y_n)^\top$, $\boldsymbol{x}_1 = (x_{11}, \ldots, x_{1n})^\top$, $\boldsymbol{x}_{1,\text{mis}} = (x_{11}, \ldots, x_{1m})^\top$, $\boldsymbol{x}_{1,\text{obs}} = (x_{1(m+1)}, \ldots, x_n)^\top$, $\boldsymbol{x}_2 = (x_{21}, \ldots, x_{2n})^\top$, and $\boldsymbol{\theta} = (\sigma, \alpha, \delta, \beta_0, \beta_1, \beta_2)^\top$. We can formulate the EM algorithm for computing the MLE of $\boldsymbol{\theta}$ and the covariance of the MLE. Although the analytic form of the information matrix may be obtained, we will estimate the covariance of the MLE using the observed Fisher information matrix instead.

In the E step, we compute

$$
\begin{aligned}
&Q(\boldsymbol{\theta}, \boldsymbol{\theta}^{(t)}) \\
&= E(l_{\text{comp}}(\boldsymbol{\theta}; \boldsymbol{y}, \boldsymbol{x}_1, \boldsymbol{x}_2) \mid \boldsymbol{y}, \boldsymbol{x}_{1,\text{obs}}, \boldsymbol{x}_2, \boldsymbol{\theta}^{(t)}) \\
&= -n \ln \sigma - b \ln \delta - \frac{1}{2\sigma^2} \sum_{i=m+1}^{n} (y_i - \beta_0 - \beta_1 x_{1i} - \beta_2 x_{2i})^2 \\
&\quad - \frac{1}{2\delta^2} \sum_{i=m+1}^{n} (x_{1i} - \alpha)^2 - E\left\{ \frac{1}{2\delta^2} \sum_{i=1}^{m} (x_{1i} - \alpha)^2 \mid \boldsymbol{y}, \boldsymbol{x}_{1,\text{obs}}, \boldsymbol{x}_2, \boldsymbol{\theta}^{(t)} \right\} \\
&\quad - E\left\{ \frac{1}{2\sigma^2} \sum_{i=1}^{m} (y_i - \beta_0 - \beta_1 x_{1i} - \beta_2 x_{2i})^2 \mid \boldsymbol{y}, \boldsymbol{x}_{1,\text{obs}}, \boldsymbol{x}_2, \boldsymbol{\theta}^{(t)} \right\}.
\end{aligned}
$$

Note that

$$
\begin{aligned}
&E\left\{ \frac{1}{2\sigma^2} \sum_{i=1}^{m} (y_i - \beta_0 - \beta_1 x_{1i} - \beta_2 x_{2i})^2 \mid \boldsymbol{y}, \boldsymbol{x}_{1,\text{obs}}, \boldsymbol{x}_2, \boldsymbol{\theta}^{(t)} \right\} \\
&= \frac{1}{2\sigma^2} \sum_{i=1}^{m} \{ (y_i - \beta_0 - \beta_2 x_{2i})^2 + \beta_1^2 E(x_{1i}^2 \mid \boldsymbol{y}, \boldsymbol{x}_{1,\text{obs}}, \boldsymbol{x}_2, \boldsymbol{\theta}^{(t)}) \\
&\qquad - 2\beta_1 (y_i - \beta_0 - \beta_2 x_{2i}) E(x_{1i} \mid \boldsymbol{y}, \boldsymbol{x}_{1,\text{obs}}, \boldsymbol{x}_2, \boldsymbol{\theta}^{(t)}) \}
\end{aligned}
$$

and

$$
\begin{aligned}
&E\left\{ \frac{1}{2\delta^2} \sum_{i=1}^{m} (x_{1i} - \alpha)^2 \mid \boldsymbol{y}, \boldsymbol{x}_{1,\text{obs}}, \boldsymbol{x}_2, \boldsymbol{\theta}^{(t)} \right\} \\
&= \frac{1}{2\delta^2} \sum_{i=1}^{m} \left\{ E(x_{1i}^2 \mid \boldsymbol{y}, \boldsymbol{x}_{1,\text{obs}}, \boldsymbol{x}_2, \boldsymbol{\theta}^{(t)}) - 2\alpha E(x_{1i} \mid \boldsymbol{y}, \boldsymbol{x}_{1,\text{obs}}, \boldsymbol{x}_2, \boldsymbol{\theta}^{(t)}) + \alpha^2 \right\}.
\end{aligned}
$$

Using the fact that if (X, Y) follows a bivariate normal distribution, $N(\mu_x, \mu_y, \sigma_X^2, \sigma_Y^2, \rho)$, then $X \mid Y \sim N(\mu_X + (Y - \mu_Y)\rho\sigma_X/\sigma_Y, \sigma_X^2(1 - \rho^2))$, we have

$$
E(x_{1i} \mid y_i, x_{2i}, \boldsymbol{\theta}^t) = \alpha^t + \frac{\beta_1^t \delta^{t2}}{\sigma^{t2} \delta^{t2}} (y_i - \beta_0^t - \beta_1^{(t)} \alpha' - \beta_2^{(t)} x_{2i})
$$

and

$$\begin{aligned}E(x_{1i}^2 \mid y_i, x_{2i}, \boldsymbol{\theta}^t) &= \mathrm{Var}(x_{1i} \mid y_i, x_{2i}, \boldsymbol{\theta}^t) + \{E(x_{1i} \mid y_i, x_{2i}, \boldsymbol{\theta}^t)\}^2 \\ &= \frac{\sigma^{t2}\delta^{t2}}{\sigma^{t2} + \beta_1^{t2}\delta^{t2}} + \{E(x_{1i} \mid y_i, x_{2i}, \boldsymbol{\theta}^t)\}^2.\end{aligned}$$

In the M step (maximization), we solve for $\boldsymbol{\theta}$ that maximizes $Q(\boldsymbol{\theta}, \boldsymbol{\theta}^{(t)})$. Note that the expectations in $Q(\boldsymbol{\theta}, \boldsymbol{\theta}^{(t)})$ can be solved by simply replacing missing x_{mis} with their first- and second-order conditional moments. When (Y, X) follows a multivariate normal distribution, using the basic probability distribution theory, the conditional distribution of $X \mid Y$ is also normal; therefore, the first- and second-order conditional expectation of $X \mid Y$ can be written in closed form, as already demonstrated. The algorithm for the multivariate normal distribution has been discussed in detail by Schafer (1997b) and Little and Rubin (2002), and has been implemented in R package `norm` (Schafer, 2013) when (X, Y) are multivariate normal.

4.3.2 EM Algorithm for Linear Regression with Missing Discrete Covariate

Here, we consider the same problem as in Section 4.3.1 except that X_1 is discrete. For missing data in multiple discrete covariates, see Ibrahim (1990) and Horton and Laird (1999) for details.

For illustrative purposes, we outline the EM algorithm for binary X_1. It can be easily extended to the case where X_1 takes more than two discrete values. Let $p_0 = \Pr(X_1 = 0 \mid X_2)$ and $p_1 = \Pr(X_1 = 1 \mid X_2) = 1 - p_0$. Note that p_0 and p_1 could be a function of X_2. For example, $p_1 = \frac{\exp(\eta_0+\eta_1 X_2)}{1+\exp(\eta_0+\eta_1 X_2)}$ if we assume a logistic regression of X_1 on X_2. We also let $\boldsymbol{x}_{\mathrm{obs}} = \left(\boldsymbol{x}_{1,\mathrm{obs}}^\top, \boldsymbol{x}_2^\top\right)^\top$. Then, we have

$$\begin{aligned}Q(\theta, \theta') &= E\left\{l_{\mathrm{comp}}(\theta; \boldsymbol{y}, \boldsymbol{x}) \mid \boldsymbol{y}, \boldsymbol{x}_{\mathrm{obs}}, \theta'\right\} \\ &= -n \log \sqrt{2\pi} - n \log \sigma - \frac{1}{2\sigma^2} \sum_{i=m+1}^{n} (y - \beta_0 - \beta_1 x_{1i} - \beta_2 x_{2i})^2 \\ &\quad + \sum_{i=m+1}^{n} \left\{(1 - x_{1i}) \log p_0 + x_{1i} \log p_1\right\} \\ &\quad - E\left\{\frac{1}{2\sigma^2} \sum_{i=1}^{m} (y - \beta_0 - \beta_1 x_{1i} - \beta_2 x_{2i})^2 \mid \boldsymbol{y}, \boldsymbol{x}_{\mathrm{obs}}, \theta'\right\} \\ &\quad + E\left\{\sum_{i=1}^{m} (1 - x_{1i}) \log p_0 + x_{1i} \log p_1 \mid \boldsymbol{y}, \boldsymbol{x}_{\mathrm{obs}}, \theta'\right\}.\end{aligned}$$

For the two expectation terms in the above formula, we have

$$
\begin{aligned}
& E\left\{\sum_{i=1}^{m}(y_i - \beta_0 - \beta_1 x_{1i} - \beta_2 x_{2i})^2 \mid \boldsymbol{y}, \boldsymbol{x}_{\mathrm{obs}}, \theta'\right\} \\
&= \sum_{i=1}^{m}\left(y_i - \beta_0 - \beta_2 x_{2i}\right)^2 + \beta_1^2 \sum_{i=1}^{m} E(x_{1i}^2 \mid y_i, x_{2i}, \theta') \\
&\quad -2\beta_1 \sum_{i=1}^{m}\left(y_i - \beta_0 - \beta_2 x_{2i}\right) E\left(x_{1i} \mid y_i, x_{2i}, \theta'\right)
\end{aligned}
$$

and

$$
\begin{aligned}
& E\left\{\sum_{i=1}^{m}(1 - x_{1i})\log p_0 + x_{1i}\log p_1) \mid \boldsymbol{y}, \boldsymbol{x}_{\mathrm{obs}}, \theta'\right\} \\
&= \log(p_0) + E(x_{1i} \mid y_i, x_{2i}, \theta')\log(p_1/p_0).
\end{aligned}
$$

We only need to calculate $E(x_{1i} \mid y_i, x_{2i}, \theta')$, since for binary data $E(x_{1i}^2 \mid y_i, x_{2i}, \theta') = E(x_{1i} \mid y_i, x_{2i}, \theta')$. Denote $P'_{ij} \stackrel{\Delta}{=} P(x_{1i} = j \mid y_i, x_{2i}, \theta')$ for $j = 0, 1$. Using Bayes' rule,

$$
P'_{ij} = \frac{p'_j \exp\left\{-(1/2\sigma'^2)(y_i - \beta'_0 - j\beta'_1 - \beta'_2 x_{2i})^2\right\}}{\sum_{k=0}^{1} p'_k \exp\left\{-(1/2\sigma'^2)(y_i - \beta'_0 - k\beta'_1 - \beta'_2 x_{2i})^2\right\}}.
$$

This is a special case of the more general weighted EM approach proposed by Ibrahim (1990). It is straightforward to extend the EM algorithm to incorporate auxiliary information; see Horton and Laird (1999) for more information. In the above algorithm, the weights are calculated for values 0 and 1 since the covariate subject to missingness is binary. If the covariate has multiple categories, the weights would be calculated for each possible category of the variable. If two or more discrete covariates have missing values, the weights should be the probability of taking on a specific combination of these covariates conditioning on current parameter estimates.

In the M step, we maximize $Q(\theta, \theta')$ with respect to θ. For standard errors and other theoretical properties of the maximum likelihood estimates, see the references mentioned above.

4.3.3 EM Algorithm for Logistic Regression with Missing Binary Outcome

Assume that we observe an independent and identically distributed (i.i.d.) sample of size n, where the binary responses $y_i \sim \text{Bernoulli}(p_i)$ are linked with the vector of covariates x_i by a logistic regression model given by

$$\log\left(\frac{p_i}{1-p_i}\right) = \beta_0 + \beta_1^{\top} x_i.$$

That is,

$$p_i = \frac{\exp(\beta_0 + \beta_1^{\top} x_i)}{1 + \exp(\beta_0 + \beta_1^{\top} x_i)}.$$

Then, the full data log-likelihood is

$$l_{\text{comp}}(\beta) = \sum_{i=1}^{n} \left\{ y_i(\beta_0 + \beta_1^{\top} x_i) - \log\left(1 + \exp(\beta_0 + \beta_1^{\top} x_i)\right) \right\}.$$

When there are missing outcomes only, the expected full data log-likelihood can be calculated by replacing the expectation with the estimate of the response probability at the current parameter values. We leave this as an exercise to readers.

4.3.4 Simulation Study

Consider a simulation study in which weights and heights of a group of women who exercise at a fitness center are measured. Let Y_1 and Y_2 denote the measured weight and height, respectively. Further assume that (Y_1, Y_2) jointly follow a bivariate normal distribution with mean (56.89, 164.71), variance (47.05, 32.03), and covariance 19.91 (correlation 0.51). These values are chosen to resemble the relationship between measured weights (kg) and heights (cm) among a group of exercising females, as reported in Fox (2008), Figure 5.1, page 86. We set that Y_1 (weight) is completely observed, whereas Y_2 (height) is missing with probability $\exp(-0.01Y_1)/\{1 + \exp(-0.01Y_1)\}$. We can see that Y_2 is missing at random (MAR) because the missingness depends on the fully observed Y_1 only. Under this setting, we generate a sample of (Y_1, Y_2) with size 300, the scatter plot of which is given in Figure 4.1.

We first consider the regression of Y_2 on Y_1, a missing outcome problem. The results of the full data analysis, complete-case analysis, and the EM algorithm are given in Table 4.1. We can see that the result of the complete-case linear regression is close to that of the full data analysis, implying that the complete-case analysis is unbiased in this case, although there is loss of efficiency due to loss of sample as indicated by the larger standard error estimates. The intercept and slope estimates of the EM algorithm are similar to the full data analysis, with improved standard errors compared to those from the complete-case analysis.

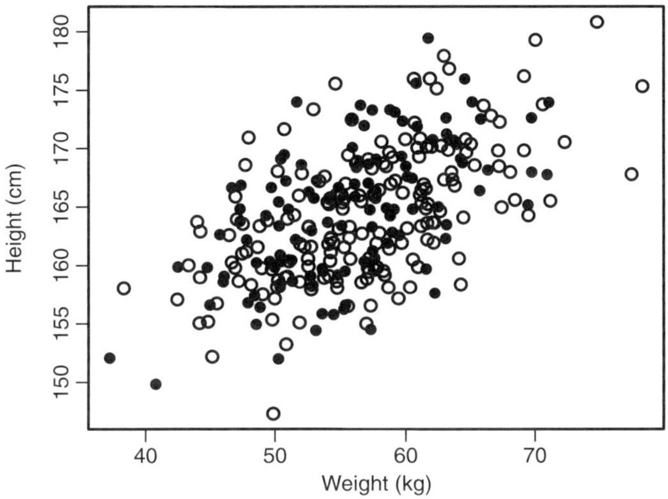

FIGURE 4.1 Scatter plot of height versus weight. The solid circles are the sample points with height observed, whereas the targets are the sample points with height missing.

TABLE 4.1 Linear Regression with Y_2 as Outcome

Coefficient	Full Data	Complete-Case	EM
Intercept	137.75 (2.26)	138.50 (2.81)	138.46 (2.20)
Weight	0.47 (0.04)	0.46 (0.05)	0.46 (0.04)

TABLE 4.2 Linear Regression with Y_2 as Covariate

Coefficient	Full Data	Complete-Case	EM
Intercept	−55.80 (9.41)	−60.39 (12.63)	−54.68 (9.69)
Height	0.68 (0.06)	0.71 (0.08)	0.68 (0.06)

Now we consider the regression of Y_1 on Y_2, a missing covariate problem. The results are summarized in Table 4.2. We can see that the EM algorithm gives a result similar to that of the full data analysis, whereas the result of the complete-case analysis is quite different from them, meaning that in this case the complete-case analysis is biased.

4.3.5 IMPACT Study

As discussed in Chapter 1, the IMPACT study is longitudinal. The primary depression outcome is measured by the Symptom Checklist-20 (SCL-20). For illustrative purposes, we conduct a cross-sectional data analysis looking at the outcomes at the sixth month. The goal is to evaluate the effect of the intervention while controlling

TABLE 4.3 Cross-Sectional Analysis Result for the IMPACT Study

	Complete-Case		EM	
	Estimate	SE	Estimate	SE
Intercept	−0.896	0.166	−0.896	0.154
Group	0.803	0.106	0.803	0.099
Baseline SCL-20	0.044	0.087	0.043	0.08

for baseline severity. We define a binary outcome, an improvement indicator, which indicates whether there is 50% or more improvement at the sixth month relative to baseline in SCL-20 scores. The group assignment and baseline SCL-20 score are used as covariates, and a logistic regression is considered.

Among the 1797 subjects with observed baseline SCL-20 scores, 1567 subjects are completely observed on the group assignments and the SCL-20 scores at baseline and the sixth month, whereas 230 subjects have their SCL-20 scores at the sixth month missing. A total of 630 subjects have their improvement indicators equal to 1, meaning that they have substantial improvement at the sixth month in SCL-20 scores.

Table 4.3 shows the results of the complete-case analysis and the EM algorithm. It can be seen that the estimates of the two methods are very close, but the estimates of the EM algorithm have smaller standard errors. Although not in all cases, in general the EM algorithm provides more efficient estimates.

4.3.6 NACC Study

For illustration, we consider a subset of 400 NACC subjects who died and have complete data on all covariates except neuropathological diagnosis of Alzheimer's disease. A linear regression model considered with continuous outcome MMSE score. The covariates are reaganAD (neuropathological, diagnosis of Alzheimer's disease), age, gender, white, married, stroke, Parkinson's disease (PD), and depression. All variables are binary except the outcome MMSE score and the covariate age. Of the 400 subjects, 273 subject are missing reaganAD. So this is a regression problem with continuous outcome and missing discrete covariate.

The regression analysis results of the complete-case analysis and the EM algorithm are shown in Table 4.4. We can see that there are large differences in coefficient estimates between the EM algorithm and the complete-case analysis.

4.4 BAYESIAN METHODS

4.4.1 Theory

In the clinical trials setting, Bayesian methods provide an approach for learning from evidence as it accumulates. In traditional (frequentist) statistical methods, information

TABLE 4.4 Cross-Sectional Analysis Result for the NACC Study

	Complete-Case		EM	
	Estimate	SE	Estimate	SE
Intercept	2.16	0.64	2.3	0.24
reaganAD	0.97	0.28	0.77	0.16
Cage	−0.15	0.14	0.04	0.08
Gender	−0.04	0.31	−0.53	0.16
White	−1.08	0.58	−0.76	0.21
Married	0.27	0.33	0.51	0.17
Stroke	0.33	0.35	0.04	0.21
PD	0.26	0.59	0.15	0.33
Depression	0.36	0.37	−0.07	0.24

from the previous studies is used only at the design stage, and never in the formal analysis of the current study. In contrast, in Bayesian methods, Bayes' theorem is used to combine prior information with current information on quantities of interest. This prior information refers to the set of all information that can be used to construct the "prior distribution," also called the "prior probabilities" or just simply the "prior".

Bayesian statistical inferences about the parameter θ of interest or missing data D_{mis} are made in terms of probability statements that are conditional on the observed value D_{obs}, or $p(\theta | D_{\text{obs}})$ and $p(D_{\text{mis}} | D_{\text{obs}})$. It is this conditioning on the observed data that makes Bayesian inference different from the traditional (frequentist) statistical inference described in most articles and textbooks, which is based on an evaluation of the procedure used to estimate θ or D_{mis} over the distribution of possible D_{obs} values conditional on the true unknown value of θ. Note that in simple settings, results from both methods can be quite similar.

4.4.2 Joint Model and Ignorable Missingness

Let D denote a rectangular data set containing values of several variables. That is, $D = (D_{ij})$, where D_{ij} is the value of variable j for subject i. We further partition D into observed and missing values, that is, $D = \{D_{\text{obs}}, D_{\text{mis}}\}$. Let $r = (r_{ij})$ be the matrix of binary indicator variables such that $r_{ij} = 1$ if D_{ij} is observed, and 0 if D_{ij} is missing. Let α and θ denote vectors of unknown parameters. Then the joint model (likelihood) of the full data is

$$f(D, r \mid \alpha, \theta) = f(D_{\text{obs}}, D_{\text{mis}}, r \mid \alpha, \theta),$$

which often cannot be evaluated in the usual way because it depends on the missing data. However, the marginal distribution of the observed data can be obtained by integrating out the missing data, that is,

$$f(D_{\text{obs}}, r \mid \alpha, \theta) = \int f(D_{\text{obs}}, D_{\text{mis}}, r \mid \alpha, \theta) \mathrm{d}D_{\text{mis}}.$$

There are two commonly used factorizations:

- *Selection Model:*

$$f(D_{\text{obs}}, D_{\text{mis}}, r \mid \alpha, \theta) = f(r \mid D_{\text{obs}}, D_{\text{mis}}, \alpha, \theta) f(D_{\text{obs}}, D_{\text{mis}} \mid \alpha, \theta).$$

- *Pattern Mixture Model:*

$$f(D_{\text{obs}}, D_{\text{mis}}, r \mid \alpha, \theta) = f(D_{\text{obs}}, D_{\text{mis}} \mid r, \alpha, \theta) f(r \mid \alpha, \theta).$$

An advantage of the selection model factorization is that it includes the *model of interest* term directly. On the other hand, the pattern mixture model corresponds more directly to what is actually observed (i.e., the distribution of the data within subgroups having different missing data patterns).

With appropriate conditional independence assumptions, the selection model can be simplified as

$$f(D_{\text{obs}}, D_{\text{mis}}, r \mid \alpha, \theta) = f(r \mid D_{\text{obs}}, D_{\text{mis}}, \alpha) f(D_{\text{obs}}, D_{\text{mis}} \mid \theta).$$

Here, $f(D_{\text{obs}}, D_{\text{mis}} \mid \theta)$ is the usual likelihood we would specify if all the data had been observed, and $f(r \mid D_{\text{obs}}, D_{\text{mis}}, \alpha)$ represents the missing data mechanism and describes the way in which the probability of an observation being missing depends on other variables (observed or not) and/or on its own values. In practice, it is a reasonable assumption that the missing data mechanism and the full data model are specified by distinct parameters. For some types of missing data, the form of the conditional distribution of r can be further simplified.

If data are missing at random, the missing data indicator depends only on the observed data, that is, $f(r \mid D_{\text{obs}}, D_{\text{mis}}, \alpha) = f(r \mid D_{\text{obs}}, \alpha)$. So,

$$\begin{aligned} f(D_{\text{obs}}, r \mid \alpha, \theta) &= f(r \mid D_{\text{obs}}, \alpha) \int f(D_{\text{obs}}, D_{\text{mis}} \mid \theta) \mathrm{d}D_{\text{mis}} \\ &= f(r \mid D_{\text{obs}}, \alpha) f(D_{\text{obs}} \mid \theta). \end{aligned}$$

If data are missing completely at random (MCAR), $f(r \mid D_{\text{obs}}, D_{\text{mis}}, \alpha) = f(r \mid \alpha)$. When α and θ are distinct, the missing indicator r contains no information about θ, and then the inference about θ can be based on $f(D_{\text{obs}} \mid \theta)$ only. The missing data mechanism is termed *ignorable* if

1. the missing data are MCAR or MAR,
2. the parameters α and θ are distinct, and
3. the priors on α and θ are independent.

Note that when the missingness is ignorable, we can ignore the model of missingness. However, it does not necessarily mean that we can ignore the missing data. In this sense, Bayesian estimation with ignorable missingness is not the same as the available data analysis.

4.4.3 Bayesian Computation for Missing Data

Bayesian analyses are often computationally intensive. However, recent advances in computational algorithms and computing speed have made it possible to carry out calculations for very complex and realistic Bayesian models, increasing the popularity of Bayesian methods (Malakoff, 1999). A basic computational tool is Markov chain Monte Carlo (MCMC) sampling, a method for simulating from the distributions of random quantities.

Algorithmically, Bayesian methods make no distinction between the missing data and the parameters, since both are stochastic (unobserved). Focus is placed on the joint distribution of the missing data and parameters conditional on the observed data, and the typical setup for the joint distribution includes a prior distribution for the parameters, a joint model for all the data (observed and missing), and a model for the missing data process (unless the missing data mechanism is ignorable, in which case this last model is unnecessary).

Missing variables are sampled from their conditional distribution via different methods, such as the Gibbs sampler. There is an imputation stage where simulated draws from the posterior predictive distribution of unobserved values D_{mis} conditional on observed values D_{obs} are imputed. Then there is a draw from the posterior distribution of the parameter θ. Inference then proceeds by averaging over the distribution of the missing values. Fully Bayesian methods for missing variables only involve incorporating an extra step in the Gibbs sampler, compared with the case of having no missing values in the data. Therefore, Bayesian methods do not require additional techniques for inference to accommodate missing data. For this reason, Bayesian methods are quite powerful.

Data augmentation (DA), an example of Gibbs sampling, is a Bayesian technique for dealing with missing data under the missing data assumption of ignorability. This procedure is based on iterating between the following two steps. We first take $t = 1$ and assume that we have a provisional first parameter value $\theta^{(1)}$.

1. *Imputation:* This is referred to as the ''I step'' and imputes the missing data by taking a random draw $D_{\text{mis}}^{(t+1)}$ from the conditional distribution $L(D_{\text{mis}} | D_{\text{obs}}, \theta^{(t)})$. The result is called the *augmented data set*.

2. *Posterior:* This is referred to as the "P step" and it takes a random draw $\theta^{(t+1)}$ from the posterior conditional distribution $L(\theta | D_{\text{obs}}, D_{\text{mis}}^{(t+1)})$ under the assumption that we have no missing data.

This procedure works well in the setting where it is easy to simulate from the posterior distribution if we had the complete data. In theory, as the number of iterations goes to infinity, the distribution of $\theta^{(t)}$ converges to the posterior distribution given only D_{obs}, and the distribution of $D_{\text{mis}}^{(t)}$ converges to the predictive distribution of D_{mis} given D_{obs}. But in practice, we must carefully monitor this convergence. This is an example of Gibbs sampling, but there are also Metropolis–Hastings versions of this procedure.

The above Bayesian model can be implemented using BUGS, WinBUGS, OpenBUGS, or JAGS. WinBUGS is freely available for Windows, OpenBUGS is freely available for both Windows and Linux systems, and JAGS (Just Another Gibbs Sampler) is freely available for Windows, Linux, and Mac OSX (see their web sites for download and installation details).

4.4.4 Simulation Example

Now we consider the data example given in Section 4.3.4. The linear regression of weight on height is considered. To carry out the Bayesian analysis, we need to specify a Bayesian model, which is given by the following BUGS code.

```
model {
  for (i in 1:N) {
    # analysis model
    Y[i] ~ dnorm(mu[i], tau);
    mu[i] <- b[1]+b[2]*X[i];
    # imputation model for missing covariate
    X[i] ~ dnorm(mx, taux);
  }
  # priors for analysis model
  b[1] ~ dunif(-1000, 1000);
  b[2] ~ dunif(-1000, 1000);
  tau ~ dgamma(1.0E-3, 1.0E-3);
  # priors for imputation model
  mx ~ dnorm(160, 0.001);
  taux ~ dgamma(0.01, 0.01);
}
```

In the above BUGS code, note that in addition to the analysis model that captures the regression relationship, the covariate with missing value is treated as a stochastic node and an imputation model is specified for it. This setup then allows missing values to be simulated from the posterior predictive distributions and uncertainty to be propagated through layers of the joint model. This is the basic idea behind the entire missing data imputation framework.

The regression coefficient estimates from the full data analysis are −55.80 for intercept and 0.68 for slope, while the corresponding posterior means of the Bayesian analysis are −53.68 and 0.67 with standard errors of 11.42 and 0.07, respectively. Results of the Bayesian analysis are very close to those from the full data analysis, but the Bayesian analysis has a larger standard error.

4.4.5 IMPACT Study

For the IMPACT study, the problem involves a regression model with missing binary outcome only. In the following BUGS code, the SCL-20 score at the sixth month is considered a stochastic node and assigned Bernoulli distribution, which then simulates the missing SCL-20 scores using Gibbs sampling.

```
model {
  for (i in 1:N) {
    # analysis model, also serves as imputation model
    # for missing outcome
    respond[i] ~ dbern(pi[i]);
    logit(pi[i]) <- b[1]+b[2]*group[i]+b[3]*scl.0[i];
  }
  b[1] ~ dunif(-1000, 1000);
  b[2] ~ dunif(-1000, 1000);
  b[3] ~ dunif(-1000, 1000);
}
```

The Bayesian estimates of the intercept and the coefficients of group assignment and baseline SCL-20 score are −0.896 (0.166), 0.805 (0.106), and 0.043 (0.087), respectively. So we can see that the results of the Bayesian analysis are quite close to those based on the complete-case analysis and EM algorithm.

4.4.6 NHANES Example

We now illustrate the full Bayesian model for missing data in both outcome and covariates, with mixed variable types using the NHANES example included in the `mice` package. There are 25 subjects in the example with outcome hyp and covariates age, bmi, and chl. The covariate age is completely observed, whereas the other three variables have missing values.

As demonstrated in the following BUGS code, in addition to the analysis model that answers the primary question about the relationship between outcome variable and covariates, there are also two carefully specified imputation models for missing outcomes and missing covariates. Investigators have to take caution to ensure that the imputation models make substantive sense and that there is additional information to be extracted from the observed data to make reasonable imputations.

```
model {
  for (i in 1:N) {
    # Analysis model, it also serves as imputation
        model
    # for missing outcome
    hyp[i] ~ dbern(pi[i]);
    logit(pi[i]) <- b[1]+b[2]*age.gp1[i]+b[3]*age.
                        gp2[i]+
                      b[4]*(bmi[i]-mean(bmi))+
                      b[3]*(chl[i]-mean(chl));

    # realistic imputation model for covariates,
    # note we can't start imputation from missing data,
    # require proper factorization
    # latent variable model can be used for discrete
        covariates
    bmi[i] ~ dnorm(mu.bmi[i], tau.bmi);
    mu.bmi[i] <- theta.bmi[1]+theta.bmi[2]*age.gp1[i]+
                 theta.bmi[3]*age.gp2[i];

    # imputation model for chl, note can't depend on
        missing hyp
    chl[i] ~ dnorm(mu.chl[i], tau.chl);
    mu.chl[i] <- theta.chl[1]+theta.chl[2]*age.gp1[i]+
                 theta.chl[3]*age.gp2[i]+
                 theta.chl[4]*(bmi[i]-mean(bmi));
  }
  # priors for analysis model
  for (j in 1:5) {
    b[j] ~ dnorm(0, 1.0E-3);
  }
  # priors for imputation models
  theta.chl[1] ~ dnorm(190, 0.001);
  for (j in 2:4) {
    theta.chl[j] ~ dnorm(0, 0.001);
  }
  theta.bmi[1] ~ dnorm(26, 0.001);
  for (j in 2:3) {
    theta.bmi[j] ~ dnorm(0, 0.001);
  }
  tau.bmi ~ dgamma(1.0E-3, 1.0E-3);
  tau.chl ~ dgamma(1.0E-3, 1.0E-3);
}
```

TABLE 4.5 Bayesian Analysis Results for the NHANES Example

	Intercept	Age Group 1	Age Group 2	bmi	chl
Mean	−1.466	−48.195	0.248	3.048	−0.135
SD	3.843	20.799	0.172	1.71	31.477

The Bayesian analysis results are given in Table 4.5. It seems that relative to age and chl, bmi has a stronger relationship with hyp.

4.5 MULTIPLE IMPUTATION

4.5.1 Theory

Instead of replacing each missing value with only one randomly imputed value, it may be more appropriate to replace each with several imputed values that reflect our uncertainty in the imputation model. This is the so-called multiple imputation. Each time we take a set of imputed values separately to form a complete data set. So if we impute five times, we will be able to form five complete data sets. Then, within each of these data sets, we perform a standard analysis (as we would have if there had been no missing values). The final step combines these inferences across data sets. In general, the MI procedure involves the following three steps:

1. *Imputation:* Create M imputations of the missing data, $D_{\text{mis}}^{(1)}, \ldots, D_{\text{mis}}^{(M)}$, under a suitable model.
2. *Analysis:* Analyze each of the M complete data sets, $(D_{\text{mis}}^{(m)}, D_{\text{obs}})$, $m = 1, \ldots, M$, in the same way.
3. *Combination:* Combine the M sets of estimates and standard errors into a single estimate and standard error using Rubin's rules (Rubin, 1987).

Suppose that we are interested in making inferences about a one-dimensional parameter θ (e.g., a coefficient from a regression model). From each of the M complete data sets, we will obtain estimates $\hat{\theta}_1, \ldots, \hat{\theta}_M$ as well as associated variances $V_1, \ldots, V_M$. Using Rubin's rules, for an overall point estimate (also referred to as the combined MI estimator), we simply average over the estimates:

$$\bar{\theta} = \frac{1}{M} \sum_{m=1}^{M} \hat{\theta}_m.$$

Then the variance of the combined MI estimator involves two components: the average within-imputation variance and the average between-imputation variance. The average within-imputation variance is defined as $\bar{V} = M^{-1} \sum_{m=1}^{M} V_m$, and the average between-imputation variance is defined as $B = [1/(M-1)] \sum_{m=1}^{M} (\hat{\theta}_m - \bar{\theta})^2$.

These are combined to form the total variance estimator:

$$T = \bar{V} + \left(1 + \frac{1}{M}\right) B.$$

For details on the theoretical justification, see Rubin (1987).

It is assumed that $(\theta - \bar{\theta})/\sqrt{T}$ follows the standard normal distribution for a large sample size and large M. But when the number of imputations M is small, the above normality assumption is questionable. Rubin (1987) and Barnard and Rubin (1999) proposed to replace the normal distribution by a t-distribution for $(\theta - \bar{\theta})/\sqrt{T}$ when M is small. That is,

$$T^{-1/2}(\theta - \bar{\theta}) \sim t_{\nu},$$

where the degrees of freedom $\nu = (M - 1)(1 + r_M^{-1})^2$. Here, $r_M = ((1 + M^{-1})B)/\bar{V}$ is the relative increase in variance due to missing data. Based on this t-distribution, a $100(1 - \alpha)\%$ confidence interval for θ is given by

$$\bar{\theta} \pm t_{\nu,1-\alpha/2}\sqrt{T},$$

and a p-value for testing the null hypothesis $\theta = \theta_0$ against a two-sided alternative is

$$2P(t_{\nu} \geq T^{-1/2} \mid \bar{\theta} - \theta_0 \mid),$$

or equivalently,

$$P(F_{1,\nu} \geq T^{-1}(\bar{\theta} - \theta)^2).$$

When the complete data sets are based on limited degrees of freedom, say υ_{com}, an additional refinement replaces ν with

$$\nu^* = (\upsilon^{-1} + \hat{\nu}_{\text{obs}}^{-1})^{-1},$$

where

$$\hat{\nu}_{\text{obs}} = (1 - r_{\text{D}})\frac{\nu_{\text{com}} + 1}{\nu_{\text{com}} + 3}\nu_{\text{com}}.$$

See Barnard and Rubin (1999) for details.

The MI method has the following advantages: (1) It allows the use of simple complete-data techniques and software once the missing data have been imputed. (2) It allows the data collector (the imputer) and the data analyst to be different. (3) It reflects the sampling variability that occurs from the fact that the values are missing. (4) It reflects uncertainty of the model if the imputations are drawn from different models. (5) It allows one set of imputations drawn in the first step to be used for many different analyses in the second step. In the above description, we did not specify in the first step of MI *how* to impute the data since there are many different methods. In Chapter 2, we learned that there are some methods for imputation that are better than others. Conventional multiple imputation imputes missing data from the posterior predictive distribution of the missing data, although there are numerous ways to build the imputation model. The conventional method of proposing a multivariate model for mixtures of categorical and continuous data can be problematic, since it is often difficult to propose a sensible joint distribution for all variables of interest. Alternative methods include the use of chained equations, also referred to as regression switching, where a plausible regression model is specified for each univariate variable that must be imputed given all other variables. There are some theoretical shortcomings to this method, since specifying the conditional distributions for each univariate variable does not guarantee a suitable joint model (van Buuren, 2013).

We recommend two approaches for imputing missing values and provide code in Section 4.9 to implement these approaches for multiple imputation. The two approaches are (1) imputation drawing from the posterior predictive distribution and (2) chained equations. There are excellent textbooks dedicated to the MI methods; see Molenberghs and Kenward (2007) and van Buuren (2013).

4.5.2 Some General Guidelines on Imputation Models and Analysis Models

The rules described in the previous section for combining complete-data inferences all assume that the sample is large enough for the usual asymptotic approximations to hold. But for smaller samples, when the asymptotic methods break down, simulation-based summaries of the posterior distribution of θ may be preferable, keeping in mind that Bayesian interpretation depends on a prior distribution.

Parametric Bayesian simulation methods depend heavily on the correct form of the parametric complete-data model. MIs created under a false model may not have a disastrous effect on the final inference, provided the analyses of imputed data sets are done under more plausible assumptions. Because the imputation and analysis steps are distinct, it is possible to have valid MI inferences even if the imputation model is different from the analysis model. The rules for combining complete-data inferences were derived under some implicit assumptions of agreement between these two models.

In the case where the analysis model is a special case of imputation model (i.e., the analyst's model is more restrictive), if the analyst's extra assumptions are true, then MI inferences will be valid. However, they may be conservative since the imputations will reflect an extra degree of uncertainty. If the analyst's extra assumptions are not true, then MI inferences will not be valid.

In the case where the analysis model is more general than the imputation model (i.e., the imputer makes assumptions on the complete data that the analyst does not), if the extra assumptions for imputation models are true, then MI inferences will be still valid. In addition, the MI estimate $\bar{\theta}$ will be more efficient than the observed data estimate derived purely from the analysis model because the MI estimate incorporates the imputer's superior knowledge about the data, a property called superefficiency. Moreover, the MI interval will have an average width that is shorter than the confidence interval derived from the observed data and the analyst's model. But if the analyst's extra assumptions are not true, then MI inferences will not be valid.

For practical considerations, the imputation model should include

- variables crucial to the analysis,
- variables that are highly predictive of the variables that are crucial to the analysis (e.g., an outcome),
- variables that are highly predictive of missingness, and
- variables that describe special features of the sample design (e.g., for probability surveys).

A general guideline is that the imputer should use a model that is general enough to preserve any associations among variables (two-, three-, or even higher way associations) that may be the target of subsequent analysis. There have been some progress in developing more formal ways of deciding which variables serves as predictors in the imputation model, see the influx and outflux statistics proposed by Van Buuren (2013).

4.5.3 Theoretical Justification for the MI Method

In this section, we outline the theoretical justification for the MI method based on a large-sample Bayesian approximation, which assumes that the observed-data posterior distribution follows a normal distribution. Due to this normal assumption, we can estimate the mean and variance with a much less number of draws needed.

If we assume that $p(\theta \mid D_{\text{obs}})$ is approximately normal, the observed-data posterior is determined by the posterior mean and variance, $E(\theta \mid D_{\text{obs}})$ and $\text{Var}(\theta \mid D_{\text{obs}})$. Note that

$$\begin{aligned} E(\theta \mid D_{\text{obs}}) &= E[E(\theta \mid D_{\text{mis}}, D_{\text{obs}}) \mid D_{\text{obs}}] \\ &= \int E\left(\theta \mid D_{\text{mis}}, D_{\text{obs}}\right) p\left(D_{\text{mis}} \mid D_{\text{obs}}\right) \mathrm{d}D_{\text{mis}}, \end{aligned}$$

where the outer expectation is taken with respect to the posterior predictive distribution, $p(D_{\text{mis}} \mid D_{\text{obs}})$. By the law of total variance, we have

$$\begin{aligned} \text{Var}(\theta \mid D_{\text{obs}}) = {} & E[\text{Var}(\theta \mid D_{\text{mis}}, D_{\text{obs}}) \mid D_{\text{obs}}] \\ & + \text{Var}[E(\theta \mid D_{\text{mis}}, D_{\text{obs}}) \mid D_{\text{obs}}]. \end{aligned}$$

Furthermore, we can show that

$$\begin{aligned} & E[\text{Var}(\theta \mid D_{\text{mis}}, D_{\text{obs}}) \mid D_{\text{obs}}] \\ & = \int \text{Var}(\theta \mid D_{\text{mis}}, D_{\text{obs}}) p(D_{\text{mis}} \mid D_{\text{obs}}) \mathrm{d}D_{\text{mis}} \end{aligned}$$

and that

$$\begin{aligned}
&\mathrm{Var}[E(\theta \mid D_{\mathrm{mis}}, D_{\mathrm{obs}}) \mid D_{\mathrm{obs}}] \\
&= \int E^2(\theta \mid D_{\mathrm{mis}}, D_{\mathrm{obs}}) p(D_{\mathrm{mis}} \mid D_{\mathrm{obs}}) \mathrm{d}D_{\mathrm{mis}} \\
&\quad - \left(\int E(\theta \mid D_{\mathrm{mis}}, D_{\mathrm{obs}}) p(D_{\mathrm{mis}} \mid D_{\mathrm{obs}}) \right)^2 .
\end{aligned}$$

Hence, if $D_{\mathrm{mis}}^{(i)}$ are independent draws of D_{mis} from the posterior predictive distribution $p(D_{\mathrm{mis}} \mid D_{\mathrm{obs}})$, for large M, we have

$$E\left[E(\theta \mid D_{\mathrm{mis}}, D_{\mathrm{obs}}) \mid D_{\mathrm{obs}}\right] \approx \frac{1}{M} \sum_{m=1}^{M} \hat{\theta}_m,$$

where $\hat{\theta}_m = E(\theta \mid D_{\mathrm{mis}}^{(m)}, D_{\mathrm{obs}})$, the complete-data posterior mean of θ calculated for the mth imputed data set $(D_{\mathrm{mis}}^{(i)}, D_{\mathrm{obs}})$. Similarly, for large M, we have

$$E[\mathrm{Var}(\theta \mid D_{\mathrm{mis}}, D_{\mathrm{obs}}) \mid D_{\mathrm{obs}}] \approx \bar{V} = \frac{1}{M} \sum_{m=1}^{M} \mathrm{Var}(\theta \mid D_{\mathrm{mis}}^{(m)}, D_{\mathrm{obs}}),$$

where $\mathrm{Var}(\theta \mid D_{\mathrm{mis}}^{(m)}, D_{\mathrm{obs}})$ is the complete-data posterior variance of θ calculated for the mth imputed data set $(D_{\mathrm{mis}}^{(m)}, D_{\mathrm{obs}})$, and

$$\mathrm{Var}[E(\theta \mid D_{\mathrm{mis}}, D_{\mathrm{obs}}) \mid D_{\mathrm{obs}}] \approx B = \frac{1}{M-1} \sum_{m=1}^{M} (\hat{\theta}_m - \bar{\theta})^2,$$

where $\bar{\theta} = M^{-1} \sum_{m=1}^{M} \hat{\theta}_m$. We call $\bar{V}$ and B the average within-imputation variance and between-imputation variance, respectively.

4.5.4 MI When θ Is k-Dimensional

When θ is not a scalar, but a vector with k dimensions, finding an adequate reference distribution for the statistic

$$(\bar{\theta} - \theta)' V(\theta \mid D_{\mathrm{obs}})^{-1} (\bar{\theta} - \theta)/k$$

is not a simple task. The main problem is that for small M, the between-imputation covariance matrix B is a poor estimate of $V(\theta \mid D_{\mathrm{obs}})$, and does not even have full rank if $M \leq k$. One solution is to make the assumption that the population between- and within-imputation covariance matrices are proportional to one another, which is equivalent to assuming that the fractions of missing information for all components of θ are equal. Under this assumption, a more stable estimator of total variance is

$$\tilde{V}(\theta \mid D_{\mathrm{obs}}) = (1 - r_M)\bar{V},$$

where $r_M = (1 + M^{-1})\mathrm{tr}(B\bar{V}^{-1})/k$ is the average relative increase in variance due to the missing data across the components of θ, and $\mathrm{tr}(B\bar{V}^{-1})$ is the trace of $B\bar{V}^{-1}$, the sum of main diagonal elements of $B\bar{V}^{-1}$.

Under the null hypothesis $H_0 : \theta = \theta_0$, the test statistic

$$W(\theta_0, \theta) = (\theta_0 - \bar{\theta})^\top \bar{V}^{-1}(\theta_0 - \bar{\theta})/(1 + r_m)k$$

follows an F-distribution with the degrees of freedom k and v_1. Hence, the p-value is $P(F_{k,v_1} > W(\theta_0, \theta))$. For $k(M-1) > 4$, the degrees of freedom are

$$v_1 = 4 + (k(M-1) - 4)\left[1 + \frac{a}{r_M}\right]^2, \quad a = 1 - \frac{2}{k(M-1)}.$$

For $k(M-1) \leq 4$, the degrees of freedom are

$$v_1 = (k+1)v/2 = (k+1)(M-1)(1 + r_M^{-1})^2/2.$$

Although the above reference distribution is derived under the strong assumption that the fractions of missing information for all components of θ are equal, Li et al. (1991) reported encouraging results even when this assumption is violated.

Assume that there are nuisance parameters ϕ, in addition to the parameter of interest θ. Our null and alternative hypotheses are $H_0 : \theta = \theta_0$ versus $H_1 : \theta \neq \theta_0$. Let $\widehat{\theta}$ and $\widehat{\phi}$ be the estimates of θ and ϕ without H_0, and let $\widehat{\phi}_0$ be the estimate of ϕ under H_0 when there are no missing data. Then, the p-value for $\theta = \theta_0$ based on the likelihood ratio test will be $\Pr(\chi_k^2 > \mathrm{LR})$, where $\mathrm{LR} = \mathrm{LR}[(\hat{\theta}, \hat{\phi}), (\theta_0, \hat{\phi}_0)]$ and χ_k^2 is a χ^2 random variable with k degrees of freedom.

For the mth imputed data set $(D_{\mathrm{mis}}^{(m)}, D_{\mathrm{obs}})$, let $(\widehat{\theta}^{(m)}, \widehat{\phi}^{(m)})$ be the estimates of θ and ϕ without assuming H_0 and $\widehat{\phi}_0$ be an estimate of ϕ under $H_0 : \theta = \theta_0$, and $\mathrm{LR}^{(i)}$ be the corresponding likelihood ratio test statistic. Let $\bar{\theta} = \sum_{m=1}^M \widehat{\theta}_m/M$, $\bar{\phi} = \sum_{m=1}^M \bar{\phi}^{(m)}$, $\bar{\phi}_0 = \sum_{m=1}^M \bar{\phi}_0^{(m)}$, and $\bar{\mathrm{LR}} = \sum_{m=1}^M \mathrm{LR}^{(m)}/M$. Assume that the function LR can be evaluated for each of the M complete data sets at $\bar{\theta}$, $\bar{\phi}$, θ_0, and $\bar{\phi}_0$ to obtain M values of $\mathrm{LR}[(\bar{\theta}, \bar{\phi}), (\theta_0, \bar{\phi}_0)]$ whose average across the m imputations is $\bar{\mathrm{LR}}_0$. Then, the test statistic

$$W = \bar{\mathrm{LR}}_0 \Big/ \left[k + \frac{(M+1)(\bar{\mathrm{LR}} - \bar{\mathrm{LR}}_0)}{(M-1)}\right]$$

has the same asymptotic distribution as $W(\theta_0, \bar{\theta})$ (Meng and Rubin, 1992). Hence, the p-value is $\Pr(F_{k,v_1} > W)$.

In some cases, the complete-data method may not produce estimates of the general function $\mathrm{LR}(\cdot)$, but only the value of the likelihood ratio statistic. So if we do not have $\bar{\mathrm{LR}}_0$ but only $\mathrm{LR}_1, \ldots, \mathrm{LR}_m$, then there is a less accurate way to combine this value (Li et al., 1991). The repeated-imputation p-value is given by

$$\Pr(F_{k,b} > \tilde{\mathrm{LR}}),$$

where

$$\tilde{\mathrm{LR}} = \frac{\bar{\mathrm{LR}}/k - (1 - m^{-1})v}{1 + (1 + M^{-1})v},$$

v is the sample variance of $(\sqrt{\mathrm{LR}_1}, \ldots, \sqrt{\mathrm{LR}_M})$, and

$$b = k^{-3/M}(M - 1)\left\{1 + [(1 + M^{-1})v]^{-1}\right\}^2.$$

4.5.5 Simulated Example

In the simulated data, weight is fully observed, whereas height contains some missing values. We carry out MI by sampling the missing values from the predictive distribution based on the complete-case regression model. We also carry out the mean imputation that imputes the missing values by the mean of the predictive distribution. Results are shown in Table 4.6. It can be seen that the mean imputation does not work well, while MI gives results similar to those of the full data analysis.

4.5.6 IMPACT Study

We carry out multiple imputation for the IMPACT study by predictive mean matching with chained equations. Note that passive imputation for the binary outcome is involved. The estimates of the intercept and the coefficients of the group assignment and baseline SCL-20 score are −0.832(0.16), 0.857(0.10), and −0.008(0.09), respectively, which are close to the results obtained by the EM algorithm and the Bayesian analysis.

TABLE 4.6 Regression Analysis for the Cross-Sectional Simulated Data Based on Imputation Methods

	Mean Imputation		MI	
	Estimate	SE	Estimate	SE
Intercept	−99.32	10.34	−56.13	10.44
Height	0.95	0.06	0.69	0.06

4.6 IMPUTING ESTIMATING EQUATIONS

The idea of imputing the estimating equations was first proposed by Paik (1997). It is also called the mean score imputation because the estimating equations are often constructed by the score functions. With the full data, suppose that we can use some estimating equations to estimate the parameter of interest θ; that is, we look for an estimate of θ by solving the following estimating equation:

$$\sum_{i=1}^{n} U(D_i;\theta) = 0,$$

where D_i is the full data of subject i. When there is possibility of missing some values of subject i, D_i can be further partitioned into the observed part $D_{\text{obs},i}$ and the missing part $D_{\text{mis},i}$. When some elements of D_i are missing, we can use the expectation of $U(D_i;\theta)$ as an "imputation" :

$$\sum_{i=1}^{n} E_{D_{\text{mis},i}\,|\,D_{\text{obs},i}} U(D_i;\theta) = 0,$$

where the expectation is taken with respect to $D_{\text{mis},i}$ given $D_{\text{obs},i}$. This estimator would be similar to the multiple imputation proposed by Wang and Robins (1998), which uses the following estimating equation:

$$\sum_{j=1}^{m}\sum_{i=1}^{n} U(D_{\text{obs},i}, \tilde{D}^{(j)}_{\text{mis},i};\theta) = 0,$$

where $\tilde{D}^{(j)}_{\text{mis},i}$ is a realization from the conditional distribution $D_{\text{mis},i}|\, D_{\text{obs},i}$. Fay (1996) proved that this estimator has the same asymptotic behavior as the traditional multiple imputation estimator. It is obvious that as m is large, $(1/m)\sum_{j=1}^{m} U(D_{\text{obs},i}, \tilde{D}^{(j)}_{\text{mis},i};\theta)$ would be close to $E_{D_{\text{mis},i}\,|\,D_{\text{obs},i}} U(D_i;\theta)$. Therefore, the imputed estimating equation can be regarded as a limiting case of the multiple imputation.

If the full data estimating equation is consistent, that is, $E\left\{U(D_i;\theta)\right\} = 0$, then the imputed estimating equation is also consistent, because

$$E\left\{E_{D_{\text{mis},i}\,|\,D_{\text{obs},i}} U(D_i;\theta)\right\} = E\left\{U(D_i;\theta)\right\} = 0.$$

In practice, an imputation model is usually needed to estimate the conditional expectation $E_{D_{\text{mis},i}\,|\,D_{\text{obs},i}}$.

4.7 INVERSE PROBABILITY WEIGHTING

4.7.1 Theory

When missing data mechanism is not missing completely at random, the subjects with complete observations often constitute a biased sample of the study population. The inverse probability weighting (IPW) method uses the idea of weighting each complete case back to the original population, and is commonly used in survey sampling. Its main advantage is its semiparametric nature and flexibility. Suppose that we observe the following data:

Group	Exposed	Non-exposed
Response	1 1 1 1	2 2 2 2

Then the average response is 1.5. If instead we only observe one subject in the exposed group and three subjects in the non-exposed group, then the average response of those observed is 7/4 = 1.75, which is biased. Note that the probability of response is 1/4 in the exposed group and 3/4 in the not exposed group. We can therefore calculate a weighted average, where each observation is weighted by the inverse of probability of being observed. Specifically, the weighted average is

$$\frac{1 \times (4/1) + (2+2+2) \times (4/3)}{(4/1) + (4/3) + (4/3) + (4/3)} = 1.5.$$

Thus, in this case, IPW has eliminated the bias through "constructing" the full population by weighting the data from subjects who have a certain probability of being observed. More generally, it can be shown that the IPW estimators are unbiased and consistent.

We now apply this idea to the problem of estimating equations with missing observations. Let R_i denote the missingness indicator, where $R_i = 1$ if subject i is completely observed, and 0 otherwise. Let $D_{\text{obs},i}$ and D_i be the observed and full data of subject i, respectively. Define $\pi(D_{\text{obs},i}) = \Pr(R_i = 1 | D_{\text{obs},i})$. The IPW estimating equation is then given as follows:

$$\sum_{i=1}^{n} \frac{R_i}{\pi(D_{\text{obs},i})} U(D_i; \theta) = 0,$$

where $E\{U(D_i;\theta)\} = 0$. After weighting each of the complete cases by the inverse of selection probability, the estimating equations would be consistent by noting that

$$\begin{aligned} E\left\{\frac{R_i}{\pi(D_{\text{obs},i})} U(D_i;\theta)\right\} &= E\left[E_{R_i|D_{\text{obs},i}}\left\{\frac{R_i}{\pi(D_{\text{obs},i})} U(D_i;\theta)\right\}\right] \\ &= E\left[\frac{\pi(D_{\text{obs},i})}{\pi(D_{\text{obs},i})} U(D_i;\theta)\right] \\ &= E\{U(D_i;\theta)\} = 0. \end{aligned}$$

In practice, the missingness mechanism $\pi(D_{\mathrm{obs},i})$ is either known by design or must be estimated from some parametric model. In the latter case, the IPW estimator requires the correct specification of the missingness mechanism model. Technically, it is also required that the selection probability is strictly positive, that is, $\pi(D_{\mathrm{obs},i}) > 0$. If some subjects are observed with very small probabilities, then the IPW estimator could be very inefficient and even unstable.

4.7.2 Simulated Example

Consider the linear regression of weight on height. The estimates of the intercept and the coefficient of height are −59.96 (12.68) and 0.71 (0.08), respectively. Compared with the complete-case analysis, we can see that the IPW substantially reduces the bias of the estimates.

4.8 DOUBLY ROBUST ESTIMATORS

4.8.1 Theory

The IPW estimator does not make full use of the available data, since the subjects with missing data are removed. Robins et al. (1994, 1995) first proposed the doubly robust (DR) estimator to improve the efficiency of the IPW estimator. Here, we focus only on the missingness of one variable. Suppose that we could partition D_i into

$$D_i = \begin{pmatrix} D_i^{(1)} \\ \\ D_i^{(2)} \end{pmatrix},$$

where only the scalar $D_i^{(1)}$ is subject to missingness and $D_i^{(2)}$ is observed for everybody. It should be noted that $D_i^{(1)}$ could be either an outcome or a covariate. The doubly robust estimator is obtained by solving the following estimating equation:

$$0 = \sum_{i=1}^{n} \left\{ \frac{R_i}{\pi(D_i^{(2)})} U(D_i;\theta) + \left(1 - \frac{R_i}{\pi(D_i^{(2)})}\right) E_{D_i^{(1)} | D_i^{(2)}} U(D_i;\theta) \right\}. \tag{4.2}$$

Usually both $\pi(D_i^{(2)})$ and $E_{D_i^{(1)} | D_i^{(2)}}(\cdot)$ need to be estimated from the data. So, the inference would incorporate the knowledge from both the missingness mechanism and the imputation model. The most attractive feature of (4.2) is that the estimating equation is consistent if either the imputation model or the missingness mechanism model is correct, but not necessarily both.

If the missingness mechanism model is correct,

$$
\begin{aligned}
& E\left\{\frac{R_i}{\pi(D_i^{(2)})}U(D_i;\theta)+\left(1-\frac{R_i}{\pi(D_i^{(2)})}\right)E_{D_i^{(1)}\mid D_i^{(2)}}U(D_i;\theta)\right\} \\
&= E\left[E_{R_i\mid D_{\text{obs},i}}\left\{\frac{R_i}{\pi(D_i^{(2)})}U(D_i;\theta)\right.\right. \\
&\qquad\left.\left.+\left(1-\frac{R_i}{\pi(D_i^{(2)})}\right)E_{D_i^{(1)}\mid D_i^{(2)}}U(D_i;\theta)\right\}\right] \\
&= E\left\{\frac{\pi(D_i^{(2)})}{\pi(D_i^{(2)})}U(D_i;\theta)+\left(1-\frac{\pi(D_i^{(2)})}{\pi(D_i^{(2)})}\right)E_{D_i^{(1)}\mid D_i^{(2)}}U(D_i;\theta)\right\} \\
&= E\left\{U(D_i;\theta)\right\}=0.
\end{aligned}
$$

If the imputation model is correct,

$$
\begin{aligned}
& E\left\{\frac{R_i}{\pi(D_i^{(2)})}U(D_i;\theta)+\left(1-\frac{R_i}{\pi(D_i^{(2)})}\right)E_{D_i^{(1)}\mid D_i^{(2)}}U(D_i;\theta)\right\} \\
&= E\left[E_{D_i^{(1)}\mid D_i^{(2)},R_i}\left\{\frac{R_i}{\pi(D_i^{(2)})}U(D_i;\theta)\right.\right. \\
&\qquad\left.\left.+\left(1-\frac{R_i}{\pi(D_i^{(2)})}\right)E_{D_i^{(1)}\mid D_i^{(2)}}U(D_i;\theta)\right\}\right] \\
&= E\left[\frac{R_i}{\pi(D_i^{(2)})}E_{D_i^{(1)}\mid D_i^{(2)}}U(D_i;\theta)+\left(1-\frac{R_i}{\pi(D_i^{(2)})}\right)E_{D_i^{(1)}\mid D_i^{(2)}}U(D_i;\theta)\right] \\
&= E\left[E_{D_i^{(1)}\mid D_i^{(2)}}U(D_i;\theta)\right]=E\left\{U(D_i;\theta)\right\}=0.
\end{aligned}
$$

When more than one variables are subject to missingness, the doubly robust estimators usually have a more complicated form. Interested readers should refer to Tsiatis (2006).

4.8.2 Variance Estimation

The imputation, IPW, and DR estimators often involve the estimation of nuisance parameters in order to evaluate $\pi(D_{\text{obs},i})$ and/or $E_{D_{\text{mis},i}\mid D_{\text{obs},i}}(\cdot)$. So before introducing the variance estimate for $\hat{\theta}$, we need to discuss the estimation of the nuisance parameters.

The estimation of $\pi(D_{\text{obs},i})$ can be obtained from a binary regression of R_i on some design matrix W_{1i}, which is expanded by the observed variables $D_{\text{obs},i}$. For example, we can assume that

$$\Pr\left(R_i = 1 \middle| D_{\text{obs},i}\right) = \text{expit}(W_{1i}^{\top}\eta_1),$$

where η_1 is the nuisance parameter in the selection model.

The estimation of $E_{D_{\text{mis},i}|D_{\text{obs},i}}(\cdot)$ is usually less straightforward. But if the full data estimating function $U(D_i;\theta)$ is linear in $D_{\text{mis},i}$, then we only need to model $E(D_{\text{mis},i}| D_{\text{obs},i})$. In the following cases, $U(D_i;\theta)$ would be linear in $D_{\text{mis},i}$: (a) missing categorical variables; or (b) missing continuous outcome. Then, we may estimate $E(D_{\text{mis},i}| D_{\text{obs},i})$ from regression models. For example, a natural choice is generalized linear models

$$E(D_{\text{mis},i}| D_{\text{obs},i}) = g(W_{2i}^{\top}\eta_2),$$

where W_{2i} is the design matrix and η_2 is the nuisance parameter in the imputation model. When $U(D_i;\theta)$ is nonlinear in $D_{\text{mis},i}$,

$$E_{D_{\text{mis},i}|D_{\text{obs},i}}(U(D_i;\theta)) = \int U(D_i;\theta) f_{D_{\text{mis},i}|D_{\text{obs},i}}\left(D_{\text{mis},i}\right) \mathrm{d}D_{\text{mis},i},$$

which requires additional steps to estimate the density function $f_{D_{\text{mis},i}|D_{\text{obs},i}}$.

Now we write $S_2(\eta) = \sum_i S_{2i}(\eta)$ to be the estimating equation for the imputation model and/or the selection model, and $S_1(\theta,\eta) = \sum_i S_{1i}(\theta,\eta)$ to be the estimating equation of the imputation, IPW, or DR estimators. Under some mild regularity conditions, $(\hat{\theta},\hat{\eta})$ follows a normal distribution asymptotically, with the sandwich-form variance:

$$\sqrt{n}\left[\begin{pmatrix}\hat{\theta}\\ \hat{\eta}\end{pmatrix} - \begin{pmatrix}\theta\\ \eta\end{pmatrix}\right] \rightarrow N(0, I^{-1}AI^{-\top}),$$

where

$$I = E\begin{pmatrix} \frac{\partial}{\partial\theta^{\top}} S_{1i}(\theta,\eta) & \frac{\partial}{\partial\eta^{\top}} S_{1i}(\theta,\eta) \\ 0 & \frac{\partial}{\partial\eta^{\top}} S_{2i}(\eta) \end{pmatrix},$$

$$A = E\begin{pmatrix} S_{1i}(\theta,\eta)S_{1i}^{\top}(\theta,\eta) & S_{1i}(\theta,\eta)S_{2i}^{\top}(\eta) \\ S_{2i}(\eta)S_{1i}^{\top}(\theta,\eta) & S_{2i}(\eta)S_{2i}^{\top}(\eta) \end{pmatrix}.$$

In practice, we evaluate I and A by replacing $E(\cdot)$ with the sample mean $(1/n)\sum_{i=1}^{n}(\cdot)$.

4.8.3 NACC Study

We fit the linear regression of the MMSE score (Y) versus the risk factors (X), true disease status (D), and the interaction between the disease status and the risk factors ($D \times X$). So the model is

$$E(Y \mid D, X) = \beta_0 + \beta_1^{\top} X + \beta_2 D + \beta_3^{\top} (D \times X).$$

We also allow the variance of the MMSE score to differ for AD and non-AD subjects:

$$\mathrm{var}(Y \mid D, X) = \gamma_0 + \gamma_1 D.$$

The MMSE score is transformed to $(30 - T)/5$ and the age is transformed to $(\text{age} - 70)/10$, so that the reported coefficients are of the appropriate magnitude. All the risk factors, as well as the MMSE score, are included in the missingness mechanism model. The imputation model also includes the quadratic term of the test score, as well as the interaction between the test and the covariates.

The results are shown in Table 4.7. The results from the three methods generally coincide with each other. The DR estimator identifies the main effects of race, clinical AD, and true disease status to be significant, indicating that these variables affect the magnitude of the test score. The *race* $\times$ D interaction is significant, whereas

TABLE 4.7 The Estimated Location and Scale Parameters with the Associated Standard Errors for the NACC Data

	DR	IPW	IB
Intercept	**1.394 (0.125)**	**1.452 (0.199)**	**1.408 (0.131)**
Age	−0.064 (0.039)	**−0.114 (0.051)**	−0.054 (0.032)
Gender	−0.049 (0.079)	0.046 (0.096)	−0.028 (0.074)
Race	**−0.303 (0.131)**	**−0.525 (0.203)**	**−0.304 (0.128)**
Marital status	0.083 (0.086)	**0.219 (0.102)**	0.031 (0.076)
Clinical AD	**1.120 (0.077)**	**1.160 (0.097)**	**1.128 (0.070)**
Stroke	0.129 (0.095)	0.100 (0.125)	0.129 (0.083)
Parkinson's	0.186 (0.138)	0.228 (0.153)	**0.292 (0.124)**
Depression	−0.007 (0.108)	0.144 (0.156)	−0.006 (0.083)
D	**0.984 (0.167)**	**1.090 (0.250)**	**0.982 (0.170)**
D × Age	0.049 (0.052)	0.085 (0.063)	0.035 (0.041)
D × Gender	−0.146 (0.099)	−0.065 (0.119)	**−0.180 (0.090)**
D × Race	**−0.322 (0.158)**	−0.339 (0.240)	**−0.330 (0.152)**
D × Marital status	0.039 (0.108)	−0.090 (0.127)	0.110 (0.094)
D × Clinical AD	−0.150 (0.106)	−0.154 (0.127)	−0.166 (0.095)
D × Stroke	−0.124 (0.120)	−0.082 (0.156)	−0.115 (0.105)
D × Parkinson's	0.074 (0.172)	0.012 (0.190)	−0.058 (0.155)
D × Depression	−0.225 (0.135)	−0.119 (0.180)	−0.153 (0.102)

gender, clinical AD, and depression have marginally insignificant interactions with D.

4.9 CODE USED IN THIS CHAPTER

4.9.1 Code Used in Section 4.3.4

The following code is used to generate a sample of (Y_1, Y_2) with size 300, where Y_1 is completely observed, and Y_2 is missing with probability $\exp(-0.01Y_1)/\{1 + \exp(-0.01Y_1)\}$:

```
> require(MASS)  # mvrnorm
> set.seed(1235813)
> N <- 300  # the sample size
> S <- matrix(c(47.05, 19.94, 19.94, 32.03), 2, 2)
> dd <- data.frame(mvrnorm(N, c(56.89, 164.71), S))
> colnames(dd) <- c("weight", "height")
> pmis <- expit(-0.01, dd$weight)
> drop <- rbinom(N, 1, pmis)  # about 36% missing
> simu <- data.frame(cbind(dd, drop))
```

The following code is used to carry out the full data analysis, complete-case analysis, and EM algorithm of regression of Y_2 on Y_1:

```
## full data analysis
> lm(height ~ weight, data = simu)
## complete-case analysis
> lm(height ~ weight, data = simu, subset = !drop)
## EM algorithm
# prepare data
> y <- simu$height
> r.y <- drop
> y[r.y == 1] <- NA
> X <- cbind(1, simu$weight)
> fit <- lm(y ~ X[, 2])
> d <- summary(fit)
> theta.hat <- c(d$sigma, d$coef[, 1])
# theta.hat is the initial values from ols estimates
# iterate over EM steps
> n.iter <- 0
> diff <- 10
> theta.hat <- rep(0.5, 3)  # starting values
> while (diff >= 1e-06 & n.iter < 20) {
+     my.em <- optim(theta.hat, method = "L-BFGS-B",
+         fn = eloglik.cy.my, gr = eloglik.cy.my.gradient,
```

```
+           hessian = TRUE, lower = c(1e-06, -Inf, -Inf),
+           upper = rep(Inf, 3), y = y,
+           r.y = r.y, X = X, theta.hat = theta.hat)
+       diff <- abs(max(my.em$par - theta.hat))
+       n.iter <- n.iter + 1
+       theta.hat <- my.em$par
+ }
> inverted <- solve(my.em$hessian)
> tval <- my.em$par/sqrt(diag(inverted))
> pval <- 2 * (1 - pt(abs(tval), nrow(X) - ncol(X)))
> results <- cbind(my.em$par, sqrt(diag(inverted)), tval, pval)
> colnames(results) <- c("Estimate", "Std. Error", "t value",
+        "Pr(>|t|)")
> rownames(results) <- c("sigma", "Intercept", "weight")
> print(results)
```

The following code is used to carry out the full data analysis, complete-case analysis, and EM algorithm for the regression of Y_1 on Y_2:

```
## Full data analysis
> lm(weight ~ height, data = simu)
## complete-case analysis
> lm(weight ~ height, data = simu, subset = !drop)
## EM algorithm
# prepare data
> y <- simu$weight
> x <- simu$height
> x[simu$drop == 1] <- NA
> X <- cbind(1, x)
> r.x <- simu$drop
> d <- summary(lm(weight ~ height, data = simu, subset = !drop))
> theta.cc <- c(d$sigma, mean(X[, 2], na.rm = T),
+     sd(X[, 2], na.rm = T), d$coef[, 1])
# theta.cc is the initial value for EM iterations
# iterate over EM steps
> n.iter <- 0
> diff <- 10
> theta.hat <- rep(0.5, 5)  # try different starting values
> while (diff > 1e-06 & n.iter < 50) {
+     my.em <- optim(theta.hat, method = "L-BFGS-B",
+         fn = eloglik.cycx.mx, hessian = TRUE,
+         lower = c(1e-06, -Inf, 1e-06, rep(-Inf, 2)),
+         upper = rep(Inf, 5), y = y, X = X, r.x = r.x,
+         mXcol = 2, theta.hat = theta.hat)
+     diff <- abs(max(my.em$par - theta.hat))
+     n.iter <- n.iter + 1
+     theta.hat <- my.em$par
```

```
+ }
> inverted <- solve(my.em$hessian)
# numeric hessian is not quite stable
> se <- sqrt(diag(inverted))
> tval <- my.em$par/se
> pval <- 2 * (1 - pt(abs(tval), nrow(X) - ncol(X)))
> results <- cbind(my.em$par, se, tval, pval)
> colnames(results) <- c("Estimate", "Std. Error", "t value",
+     "Pr(>|t|)")
> rownames(results) <- c("sigma", "m.x", "sd.x", "Intercept",
+     "height")
```

4.9.2 Code Used in Section 4.3.5

The following code is used to carry out the complete-case analysis and the EM algorithm for the IMPACT data:

```
## complete-case analysis
> y <- impact$respond
> r.y <- is.na(y)
> y[r.y == 1] <- NA
> X <- cbind(1, impact$group, impact$scl.0)
> d <- summary(glm(respond ~ group + scl.0, data = impact,
+    family = binomial(link = "logit"), na.action = "na.omit"))
## iterate over EM steps
> n.iter <- 0
> diff <- 10
> theta.hat <- rep(0.5, 3)
# the above code is to try different starting values
> while (diff > 1e-06 & n.iter < 50) {
+     my.em <- optim(theta.hat, method = "L-BFGS-B",
+         fn = eloglik.dy.my, hessian = TRUE,
+         lower = rep(-Inf, 3), upper = rep(Inf, 3),
+         y = y, r.y = r.y, X = X, theta.hat = theta.hat)
+     diff <- abs(max(my.em$par - theta.hat))
+     n.iter <- n.iter + 1
+     theta.hat <- my.em$par
+ }
> inverted <- solve(my.em$hessian)
> tval <- my.em$par/sqrt(diag(inverted))
> pval <- 2 * (1 - pt(abs(tval), nrow(X) - ncol(X)))
> results <- cbind(my.em$par, sqrt(diag(inverted)), tval, pval)
> colnames(results) <- c("Estimate", "Std. Error", "t value",
+     "Pr(>|t|)")
> rownames(results) <- c("Intercept", "group1", "scl.0")
```

4.9.3 Code Used in Section 4.4.4

The following code is used to carry out the Bayesian analysis for the simulation example:

```
# THE DATA
> X <- c(simu$height)  # get rid of attributes
> Y <- c(simu$weight)
> X[simu$drop == 1] <- NA
> N <- length(Y)
> dataList <- list(X = X, Y = Y, N = N)
> ii <- min(c(1:N)[is.na(X)])  # first missing value to monitor

# THE ARGUMENTS
  # The parameters to be monitored
> parameters <- c("b[1]", "b[2]", paste0("X[", ii, "]"))
  # Number of steps to 'tune' the samplers.
> adaptSteps <- 1000
  # Number of steps to 'burn-in' the samplers.
> burnInSteps <- 5000
  # Number of chains to run.
> nChains <- 3
  # Total number of steps in chains to save.
> numSavedSteps <- 10000
  # Number of steps to 'thin' (1=keep every step).
> thinSteps <- 10
  # Steps per chain.
> nIter <- ceiling((numSavedSteps * thinSteps)/nChains)

# Require R package rjags/runjags and JAGS package!
require(rjags)

# export model specification to an external text file
> writeLines(modelString, con = "mymodel.bug")

# Create, initialize, and adapt the model:
> jagsModel <- jags.model("mymodel.bug", data = dataList,
+       n.chains = nChains, n.adapt = adaptSteps)

# Burn-in:
> cat("Burning in the MCMC chain...\n")
> update(jagsModel, n.iter = burnInSteps, progress.bar = "none")

# The saved MCMC chain:
> cat("Sampling final MCMC chain...\n")
> codaSamples <- coda.samples(jagsModel, variable.names
+     = parameters, n.iter = nIter, thin = thinSteps)

# EXAMINE THE RESULTS.
```

```
> checkConvergence <- T
> if (checkConvergence) {
+     show(summary(codaSamples))
+ }
```

4.9.4 Code Used in Section 4.4.5

The following code is used to carry out the Bayesian analysis for the IMPACT study:

```
# THE DATA
> respond <- c(impact$respond)  # get rid of attributes
> group <- c(impact$group)
> scl.0 <- c(impact$scl.0)
> N <- length(respond)
> dataList <- list(respond = respond, group = group,
+     scl.0 = scl.0, N = N)
  # first missing value to monitor
> ii <- min(c(1:N)[is.na(respond)])

# THE ARGUMENTS
  # The parameters to be monitored
> parameters <- c("b[1]", "b[2]", "b[3]",
+     paste0("respond[", ii, "]"))
  # Number of steps to 'tune' the samplers.
> adaptSteps <- 1000
  # Number of steps to 'burn-in' the samplers.
> burnInSteps <- 5000
> nChains <- 3  # Number of chains to run.
  # Total number of steps in chains to save.
> numSavedSteps <- 5000
  # Number of steps to 'thin' (1=keep every step).
> thinSteps <- 10
  # Steps per chain.
> nIter <- ceiling((numSavedSteps * thinSteps)/nChains)

> require(rjags)

# export BUGS model to an external text file
> writeLines(modelString, con = "mymodel.bug")

# Create, initialize, and adapt the model:
> jagsModel <- jags.model("mymodel.bug", data = dataList,
+     n.chains = nChains, n.adapt = adaptSteps)

# Burn-in:
> cat("Burning in the MCMC chain...\n")
> update(jagsModel, n.iter = burnInSteps, progress.bar = "none")
```

```
# The saved MCMC chain:
> cat("Sampling final MCMC chain...\n")
> codaSamples <- coda.samples(jagsModel,
+     variable.names = parameters,
+     n.iter = nIter, thin = thinSteps)
```

4.9.5 Code Used in Section 4.4.6

The following code is used to carry out the Bayesian analysis for the NHANES example:

```
# THE DATA
> age.gp1 <- ifelse(c(nhanes2$age) == 1, 1, 0)
> age.gp2 <- ifelse(c(nhanes2$age) == 2, 1, 0)
  # change to 0/1, and get rid of attributes
> hyp <- c(nhanes2$hyp) - 1
> bmi <- c(nhanes2$bmi)
> chl <- c(nhanes2$chl)
> N <- length(hyp)
> dataList <- list(hyp = hyp, age.gp1 = age.gp1,
+     age.gp2 = age.gp2, bmi = bmi, chl = chl, N = N)

# THE ARGUMENTS
  # The parameters to be monitored
> parameters <- c("b[1]", "b[2]", "b[3]", "b[4]", "b[5]")
  # Number of steps to 'tune' the samplers.
> adaptSteps <- 1000
  # Number of steps to 'burn-in' the samplers.
> burnInSteps <- 5000
> nChains <- 3  # Number of chains to run.
  # Total number of steps in chains to save.
> numSavedSteps <- 5000
  # Number of steps to 'thin' (1=keep every step).
> thinSteps <- 10
  # Steps per chain.
> nIter <- ceiling((numSavedSteps * thinSteps)/nChains)

# Export bugs model to external text file
> writeLines(modelString, con = "mymodel.bug")

# Create, initialize, and adapt the model:
> jagsModel <- jags.model("mymodel.bug", data = dataList,
+     n.chains = nChains, n.adapt = adaptSteps)

# Burn-in:
> cat("Burning in the MCMC chain...\n")
> update(jagsModel, n.iter = burnInSteps, progress.bar = "none")
```

```
# The saved MCMC chain:
> cat("Sampling final MCMC chain...\n")
> codaSamples <- coda.samples(jagsModel,
+     variable.names = parameters,
+     n.iter = nIter, thin = thinSteps)
```

4.9.6 Code Used in Section 4.5.5

The following code is used to carry out the analysis based on the mean imputation and the multiple imputation:

```
# fit a model on height using available data
> impmodel <- lm(height ~ weight, data = simu, subset = !drop)
# replace missing height with predicted values
> simu$height.imp <- ifelse(simu$drop,
+     predict(impmodel, newdata = simu), simu$height)
# multiple imputation
> dd1 <- simu
> sigma <- summary(impmodel)$sigma
  # s is residual standard error from impmodel
> dd1$height <- ifelse(drop, rnorm(nrow(simu),
+      m = predict(impmodel, newdata = simu), s = sigma),
          dd$height)
> dd2 <- simu
> dd2$height <- ifelse(drop, rnorm(nrow(simu),
+     m = predict(impmodel, newdata = simu), s = sigma),
          dd$height)
> dd3 <- simu
> dd3$height <- ifelse(drop, rnorm(nrow(simu),
+     m = predict(impmodel, newdata = simu), s = sigma),
          dd$height)
> dd4 <- simu
> dd4$height <- ifelse(drop, rnorm(nrow(simu),
+     m = predict(impmodel, newdata = simu), s = sigma),
          dd$height)
> dd5 <- simu
> dd5$height <- ifelse(drop, rnorm(nrow(simu),
+     m = predict(impmodel, newdata = simu), s = sigma),
          dd$height)
> library(mitools)  # to combine imputed data
> ddimp <- imputationList(list(dd1, dd2, dd3, dd4, dd5))
> models <- with(ddimp, lm(weight ~ height))
```

4.9.7 Code Used in Section 4.5.6

The following code is used to multiply impute missing value using predictive mean matching, involving passive imputation:

```
> impact.imp <- mice(impact, method = c(rep("pmm", 4),
+     "~ifelse((scl.6-scl.0) <= -scl.0/2, 1, 0",
+     "~scale(scl.6)"), printFlag = FALSE)
> fit <- with(impact.imp, glm(respond ~ group + scl.0,
+     family = binomial))
> pool(fit)
```

4.9.8 Code Used in Section 4.7.2

The following code is used to carry out the IPW estimation for the simulated data example:

```
# estimate the probability of being observed
> pmodel <- glm(drop ~ weight, data = simu, family = binomial)
> estp <- 1 - fitted(pmodel)
# weighted results
> s <- lm(weight ~ height, data = simu, subset = !drop,
+     weights = 1/estp)
```

CHAPTER 5

LONGITUDINAL DATA METHODS

5.1 OVERVIEW

In contrast to the cross-sectional studies discussed in Chapter 4, longitudinal studies have the defining feature of repeated measures collected on individuals over time, enabling a direct study of temporal patterns or trajectories. Although both cross-sectional and longitudinal studies can look at differences among individuals in their baseline values (called *cohort effects* in population studies), only a longitudinal study can look at changes over time within an individual (called *aging effects* in population studies). Longitudinal data can be collected prospectively, following individuals forward in time, or retrospectively, looking back at historical records. The methods described in this chapter can be used for either data collection method.

We begin this chapter by introducing two data examples in Section 5.2 and carry out some basic descriptive analysis. Section 5.3 reviews the modeling approaches and statistical inferences for longitudinal data without missing values. We then introduce the settings of missing longitudinal data as well as simple methods to deal with missingness in Section 5.4. When only the response variable is subject to monotone missingness (e.g., dropout), Section 5.5 presents the likelihood-based method and Section 5.6 describes the inverse probability weighted generalized estimating equation approaches. Section 5.7 extends the WEE to the situation of intermittent

Applied Missing Data Analysis in Health Sciences, First Edition. Xiao-Hua Zhou, Chuan Zhou, Danping Liu, and Xiaobo Ding.

missingness of the outcome. The multiple imputation (MI) procedures are introduced in Section 5.8. Section 5.9 presents the Bayesian inference framework in the setting where both outcomes and covariates are subject to missingness. Finally, other existing methods in analyzing longitudinal missing data are introduced in Section 5.10. Technical details of some methods and the code we used in this chapter are presented in Appendix 5.A.

5.2 EXAMPLES

5.2.1 IMPACT Study

The Improving Mood and Promoting Access to Collaborative Treatment (IMPACT) study has been introduced in Chapters 1 and 4. Briefly, to determine the effectiveness of the IMPACT collaborative care management program for late life depression, a total of 1801 patients aged 60 years or older with major depression (17%), dysthymic disorder (30%), or both (53%) were enrolled and randomized to IMPACT care or usual care. Intervention patients had access for up to 12 months to a depression care manager who was supervised by a psychiatrist and a primary care expert and who offered education, care management, and support of antidepressant management by the patient's primary care physician or a brief psychotherapy for depression. Depression symptoms were assessed by the mean score of the 20 depression items from the Symptom Checklist-90 (SCL-20) at baseline, 3, 6, 12, 18, and 24 months. For details about the study design and primary analysis, see Unützer et al. (2002).

In the following, we conduct longitudinal data analysis using the outcome, the SCL-20 depression score, and contrast various missing data methods. Figure 5.1 shows trajectories of 40 randomly selected individuals and the group mean trajectories over time. It is pretty obvious that individuals from the intervention arm tend to have lower SCL-20 scores.

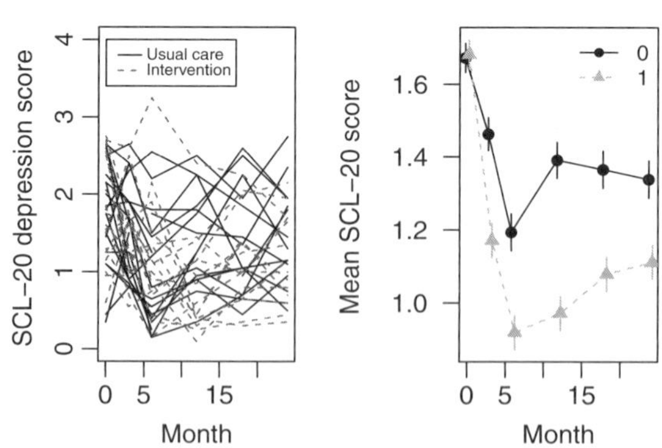

FIGURE 5.1 Forty individuals and mean trajectories.

We consider an analysis that addresses the question whether or not the IMPACT treatment reduces the depression symptoms as measured by the SCL-20 scores. To address this question, we compare the subject-specific changes, from baseline to follow-up, in the SCL-20 scores for the patients in the two study arms. We consider the following linear mixed effects regression model for the subject-specific mean SCL-20 scores:

$$\begin{aligned} E(Y_{ij} \mid b_i) &= \beta_1 + \beta_2\mathbf{period}_{ij} + \beta_3\mathbf{group}_i + \beta_4\mathbf{group}_i \times \mathbf{period}_{ij} \\ &\quad + b_{1i} + b_{2i}\mathbf{period}_{ij} \\ &= (\beta_1 + b_{1i}) + (\beta_2 + b_{2i})\mathbf{period}_{ij} + \beta_3\mathbf{group}_i \\ &\quad + \beta_4\mathbf{group}_i \times \mathbf{period}_{ij}, \end{aligned} \tag{5.1}$$

where Y_{ij} is the SCL-20 score for the ith patient at the jth time point of assessment ($j = 0, 3, 6, 12, 18, 24$). The variable **group** is an indicator variable for the treatment group, with **group** $= 0$ if an individual was randomized to the usual care group and **group** $= 1$ if randomized to the IMPACT treatment group. The binary variable **period** denotes the baseline and follow-up periods, with **period** $= 0$ for the baseline and **period** $= 1$ for the follow-up periods (3–24 months). This allows us to assess the time-averaged intervention effect. We can also replace **period** with the actual **month** variable, which then allows us to assess treatment effect at specific months. Although the SCL-20 score is bounded between 0 and 4, because it is an average of 20 items, we assume that it follows a normal distribution in our analysis. Finally, we assume that the random intercepts and slopes, b_i, follow a bivariate normal distribution, with zero mean and an unstructured 2×2 covariance matrix G.

We decided to illustrate our analysis using the Stata software version 13.0. The point is that many missing data methods have been implemented in various software packages, and the readers can choose the ones that they are already familiar with or more comfortable using.

The misstable command in Stata allows us to examine the missing data pattern in SCL-20 scores, which is shown in Table 5.1. It can be seen that 162 (9%) subjects had intermittent missingness, 370 (20%) subjects dropped out during the 24-month period, and 1269 (71%) patients had complete data.

5.2.2 NACC UDS Data

The National Alzheimer's Coordinating Center (NACC) Uniform Data Set has already been introduced in Chapter 1. The data set was also analyzed by Monsell et al. (2012). Our aim is to model the decline of the outcome, mini-mental state examination (MMSE), since the first diagnosis of amnestic mild cognitive impairment (aMCI).

TABLE 5.1 The Missing Data Pattern in the SCL-20 Scores at Baseline, 3, 6, 12, 18, and 24 Months

Frequency	SCL0	SCL3	SCL6	SCL12	SCL18	SCL24
1269	1	1	1	1	1	1
117	1	0	0	0	0	0
69	1	1	0	0	0	0
64	1	1	1	0	0	0
60	1	1	1	1	0	0
60	1	1	1	1	1	0
45	1	0	1	1	1	1
22	1	1	1	1	0	1
14	1	1	1	0	1	1
12	1	0	1	0	0	0
11	1	1	0	1	1	1
5	1	0	0	1	1	1
5	1	1	0	1	0	0
4	1	0	0	1	0	0
4	1	0	1	1	1	0
4	1	1	0	1	0	1
4	1	1	1	0	0	1
4	1	1	1	0	1	0
3	1	0	0	0	1	0
3	1	0	1	0	0	1
3	1	0	1	1	0	0
3	1	1	0	0	0	1
2	0	1	1	1	1	1
2	1	0	0	0	1	1
2	1	0	0	1	0	1
2	1	0	1	0	1	1
2	1	1	0	0	1	1
2	1	1	0	1	1	0
1	0	0	0	0	0	0
1	0	1	1	1	1	0
1	1	0	1	1	0	1
1	1	1	0	0	1	0

The total number of patients in the study is 1801.

Missing data occur in the outcome MMSE as well as in the covariates education, Hachinski ischemic score (HIS), and apolipoprotein E (APOE) e4 alleles. The missing pattern is nonmonotone, as shown in Table 5.2.

The data set is in a "long format," meaning that each row corresponds to one person-visit, and each subject contributes multiple rows. We observe that HIS was missing for 300 clinical visits and MMSE for 404 visits. Education and APOE e4 are on the subject level; it turns out that 4 subjects had missing education and 949 subjects had missing APOE e4.

TABLE 5.2 Missing Data Pattern for the UDS Data

Frequency (by Visits)	MMSE	Education	HIS	APOE e4
5809	1	1	1	1
98	0	1	1	1
13	1	0	1	1
23	1	1	0	1
2482	1	1	1	0
1	0	0	1	1
156	0	1	0	1
49	0	1	1	0
21	1	1	0	0
100	0	1	0	0

5.3 LONGITUDINAL REGRESSION MODELS FOR COMPLETE DATA

The analysis of longitudinal data requires us to account for the correlation structure among observations of the same individual measured repeatedly over time. There are three major categories of methods for analyzing longitudinal data: *marginal models*, *random effects models*, and *transition models*. In this chapter, we briefly review the first two methods because they are relatively easy to implement in many software packages, and most missing data methodologies are based on the first two methods. More detailed description of the longitudinal data analysis can be found in Diggle et al. (2002).

Suppose that N individuals are independently observed in the sample. Let Y_{ij} be the response for ith subject measured at time t_{ij}, where $i = 1, \ldots, N$ and $j = 1, \ldots, n_i$. Note that the number of observations is allowed to be different across the individuals, which is also known as an *unbalanced design*. We write $\boldsymbol{Y}_i = (Y_{i1}, Y_{i2}, \ldots, Y_{in_i})'$ to denote the $n_i \times 1$ response vector for subject i that may be missing at some time points. Let $\boldsymbol{X}_i = (X_{i1}, X_{i2}, \ldots, X_{in_i})'$ be an $n_i \times q$ matrix of covariates for subject i at each time point. The jth row of $\boldsymbol{X}_i$, X'_{ij}, corresponds to the q covariates observed at time t_{ij}. For now, we do not distinguish the covariates that change over time (e.g., age, longitudinal MMSE score measured at each t_{ij}) from those that do not (e.g., gender, baseline MMSE score measured at t_{i0}).

5.3.1 Linear Mixed Models for Continuous Longitudinal Data

5.3.1.1 Model Setting Suppose the outcome vector $\boldsymbol{Y}_i$ takes continuous values, so that we could consider a linear regression model. A linear mixed model (LMM) could be illustrated from a two-stage model formulation. We first fit a regression line for each subject separately:

$$\boldsymbol{Y}_i = \boldsymbol{Z}_i \boldsymbol{\beta}_i + \boldsymbol{\varepsilon}_i, \tag{5.2}$$

where $\boldsymbol{\beta}_i$ is a q-dimensional vector of subject-specific regression coefficients, $\boldsymbol{Z}_i$ is the design matrix of covariates, and $\boldsymbol{\varepsilon}_i \sim N(\mathbf{0}, \Sigma_i)$ is the Gaussian residual. Note that $\boldsymbol{\varepsilon}_i$ describes the within-subject variability, and usually we assume that the elements of $\boldsymbol{\varepsilon}_i$ are independent, that is, $\Sigma_i = \sigma^2 I_{n_i}$. The second-stage model imposes a structure on $\boldsymbol{\beta}_i$:

$$\boldsymbol{\beta}_i = K_i \boldsymbol{\beta} + \boldsymbol{b}_i, \tag{5.3}$$

where K_i is the design matrix, $\boldsymbol{\beta}$ is the vector of regression parameters of interest, and $\boldsymbol{b}_i \sim N(\mathbf{0}, D)$ is the residual. Note that $\boldsymbol{b}_i$ represents the variability between subjects, whose elements are usually correlated. We could regard $\boldsymbol{\beta}$ as the "average" covariates effect, and then $\boldsymbol{b}_i$ is the effect deviation from the population average for subject i.

The two-stage model, defined by (5.2) and (5.3), could be combined into the following one-stage model:

$$\boldsymbol{Y}_i = \left(\boldsymbol{Z}_i K_i\right) \boldsymbol{\beta} + \boldsymbol{Z}_i \boldsymbol{b}_i + \boldsymbol{\varepsilon}_i.$$

We call $\boldsymbol{\beta}$ the *fixed effects* and $\boldsymbol{b}_i$ the *random effects* due to the fact that $\boldsymbol{\beta}$ is the unknown parameter vector, whereas $\boldsymbol{b}_i$ is a random vector. More generally, the following four assumptions constitute an LMM:

$$\begin{aligned}
&\boldsymbol{Y}_i = \boldsymbol{X}_i \boldsymbol{\beta} + \boldsymbol{Z}_i \boldsymbol{b}_i + \boldsymbol{\varepsilon}_i,\\
&\boldsymbol{b}_i \sim N(\mathbf{0}, D(\boldsymbol{\alpha})),\\
&\boldsymbol{\varepsilon}_i \sim N(\mathbf{0}, \Sigma_i(\boldsymbol{\alpha})),\\
&\boldsymbol{b}_1, \ldots, \boldsymbol{b}_N, \boldsymbol{\varepsilon}_1, \ldots, \boldsymbol{\varepsilon}_N \text{ are independent,}
\end{aligned}$$

where $D(\boldsymbol{\alpha})$ and $\Sigma_i(\boldsymbol{\alpha})$ are called the *variance components* and $\boldsymbol{\alpha}$ is a vector of nuisance parameters specifying these variance components. The LMM can be equivalently viewed as a hierarchical model; that is, a conditional model for $\boldsymbol{Y}_i$ given $\boldsymbol{b}_i$ and a marginal model for $\boldsymbol{b}_i$:

$$\boldsymbol{Y}_i | \boldsymbol{b}_i \sim N(\boldsymbol{X}_i \boldsymbol{\beta} + \boldsymbol{Z}_i \boldsymbol{b}_i, \Sigma_i(\boldsymbol{\alpha})), \tag{5.4}$$

$$\boldsymbol{b}_i \sim N(\mathbf{0}, D(\boldsymbol{\alpha})). \tag{5.5}$$

Marginally, $\boldsymbol{Y}_i$ also follows a multivariate normal (MVN) distribution:

$$\boldsymbol{Y}_i \sim N(\boldsymbol{X}_i \boldsymbol{\beta}, \boldsymbol{Z}_i D(\boldsymbol{\alpha}) \boldsymbol{Z}_i' + \Sigma_i(\boldsymbol{\alpha})). \tag{5.6}$$

The LMM implies the marginal model, but the converse is not true. So, the parameter vector $\boldsymbol{\beta}$ may have two interpretations: the population-average covariate effects, or the covariate effects conditional on a given subject. The compatibility of the marginal and conditional model interpretation holds only for the LMM, not the generalized linear mixed models (GLMM), as we will see in the later sections.

5.3.1.2 *Model Estimation and Inference* We first assume that the vector of nuisance parameters in the variance components, $\boldsymbol{\alpha}$, is known, and write $V_i(\boldsymbol{\alpha}) = \boldsymbol{Z}_i D(\boldsymbol{\alpha}) \boldsymbol{Z}_i' + \Sigma_i(\boldsymbol{\alpha})$. Now $\boldsymbol{\beta}$ could be solved from weighted least squares:

$$\widehat{\boldsymbol{\beta}}(\boldsymbol{\alpha}) = \left(\sum_{i=1}^{N} \boldsymbol{X}_i' V_i^{-1}(\boldsymbol{\alpha}) \boldsymbol{X}_i \right)^{-1} \sum_{i=1}^{N} \boldsymbol{X}_i' V_i^{-1}(\boldsymbol{\alpha}) \boldsymbol{Y}_i. \tag{5.7}$$

It is easy to verify that (5.7) is the maximum likelihood estimator (MLE) since the likelihood function is

$$L(\boldsymbol{\alpha}, \boldsymbol{\beta}) \propto \prod_{i=1}^{N} |V_i(\boldsymbol{\alpha})|^{-1/2} \exp\left[-\frac{1}{2} \left(\boldsymbol{Y}_i - \boldsymbol{X}_i\boldsymbol{\beta}\right)' V_i^{-1}(\boldsymbol{\alpha}) \left(\boldsymbol{Y}_i - \boldsymbol{X}_i\boldsymbol{\beta}\right) \right]. \tag{5.8}$$

We should note that the likelihood inference utilizes only the marginal model (5.6), but not the hierarchical models (5.4) and (5.5).

In most cases, however, $\boldsymbol{\alpha}$ is unknown and needs to be estimated. Two common methods for estimating $\boldsymbol{\alpha}$ are maximum likelihood (ML) and restricted maximum likelihood (REML). The ML method estimates $\boldsymbol{\alpha}$ and $\boldsymbol{\beta}$ simultaneously by maximizing the likelihood function (5.8). The score equations can be derived as follows:

$$\begin{aligned} U_{\beta}(\boldsymbol{\alpha}, \boldsymbol{\beta}) &= -\sum_{i=1}^{N} \boldsymbol{X}_i' V_i^{-1}(\boldsymbol{\alpha}) \left(\boldsymbol{Y}_i - \boldsymbol{X}_i\boldsymbol{\beta}\right), \\ U_{\alpha_j}(\boldsymbol{\alpha}, \boldsymbol{\beta}) &= \sum_{i=1}^{N} \frac{1}{2} \left[\left(\boldsymbol{Y}_i - \boldsymbol{X}_i\boldsymbol{\beta}\right)' V_i^{-1}(\boldsymbol{\alpha}) \frac{\partial V_i(\boldsymbol{\alpha})}{\partial \alpha_j} V_i^{-1}(\boldsymbol{\alpha}) \left(\boldsymbol{Y}_i - \boldsymbol{X}_i\boldsymbol{\beta}\right) \right. \\ &\qquad \left. - \operatorname{tr}\left(V_i^{-1}(\boldsymbol{\alpha}) \frac{\partial V_i(\boldsymbol{\alpha})}{\partial \alpha_j} \right) \right], \end{aligned}$$

where α_j is the jth element of $\boldsymbol{\alpha}$ and $\operatorname{tr}(A)$ is the trace of matrix A.

The ML estimation may result in a substantial amount of bias for $\boldsymbol{\alpha}$ in small samples, due to the estimation of $\boldsymbol{\beta}$. A simple analogy is that the ML estimator $N^{-1} \sum_{i=1}^{N} \left(X_i - \bar{X}\right)^2$ is a biased estimation of the population variance. Bias correction involves replacing N^{-1} by $(N-1)^{-1}$. The REML approach uses a similar technique, and gives a better estimation of the variance components. The idea is to construct a likelihood function that is no longer a function of $\boldsymbol{\beta}$. Note that $\widehat{\boldsymbol{\beta}}(\boldsymbol{\alpha})$ in (5.7) is the sufficient statistic for $\boldsymbol{\beta}$, and REML is indeed a conditional likelihood function $L(\boldsymbol{Y}|\widehat{\boldsymbol{\beta}}(\boldsymbol{\alpha}))$. We give the form of the restricted likelihood as follows, and more details and derivations can be found in Patterson and Thompson (1971):

$$L_{\text{REML}}(\boldsymbol{\alpha}) \propto \prod_{i=1}^{N} \left| \boldsymbol{X}_i' V_i^{-1}(\boldsymbol{\alpha}) \boldsymbol{X}_i \right|^{-1/2} |V_i(\boldsymbol{\alpha})|^{-1/2}$$
$$\times \exp\left[-\frac{1}{2} \left(\boldsymbol{Y}_i - \boldsymbol{X}_i \widehat{\boldsymbol{\beta}}(\boldsymbol{\alpha}) \right)' V_i^{-1}(\boldsymbol{\alpha}) \left(\boldsymbol{Y}_i - \boldsymbol{X}_i \widehat{\boldsymbol{\beta}}(\boldsymbol{\alpha}) \right) \right].$$

Remark 5.1 *The REML estimate of the variance components is always larger in absolute value than the ML estimate (Verbeke and Molenberghs, 2000). The difference is more pronounced as the dimension of $\boldsymbol{\beta}$ gets larger or the sample size N gets smaller. In practice, if the variance components are of interest, then the REML should be used to correct for the bias. The trade-off to use the REML is that the estimation of the fixed effects is not as efficient as the ML method.*

Remark 5.2 *For testing the variance components, the likelihood ratio test is valid with either the likelihood or the restricted likelihood. But to test the fixed effects, one can never use the restricted likelihood because it is not a likelihood function for $\boldsymbol{\beta}$.*

With $\boldsymbol{\alpha}$ estimated using the ML or REML method, $\boldsymbol{\beta}$ is estimated by plugging $\widehat{\boldsymbol{\alpha}}$ into (5.7). Conditional on $\boldsymbol{\alpha}$, $\widehat{\boldsymbol{\beta}}(\boldsymbol{\alpha})$ follows a multivariate normal distribution with mean $\boldsymbol{\beta}$ and variance

$$\text{var}(\widehat{\boldsymbol{\beta}}) = \left(\sum_{i=1}^{N} \boldsymbol{X}_i' V_i^{-1} \boldsymbol{X}_i \right)^{-1}.$$

In practice, we again replace $\boldsymbol{\alpha}$ by its ML or REML estimate to evaluate this variance.

5.3.1.3 Empirical Bayes Estimates for Random Effects In some situations, we may wish to know the value of the random effects $\boldsymbol{b}_i$, so that the individual trajectories could be portrayed. Since $\boldsymbol{b}_i$ is random, the Bayesian idea is a natural choice in order to "guess" the most plausible value of a random variable. If we can find the posterior distribution $[\boldsymbol{b}_i | \boldsymbol{Y}_i, \boldsymbol{\alpha}, \boldsymbol{\beta}]$, the random effects would be easily estimated by the posterior mean $E[\boldsymbol{b}_i | \boldsymbol{Y}_i, \widehat{\boldsymbol{\alpha}}, \widehat{\boldsymbol{\beta}}]$, with the parameters replaced by the ML or REML estimates. The estimation of $\boldsymbol{b}_i$ relies on the hierarchical model, and we have

$$\begin{pmatrix} \boldsymbol{Y}_i \\ \boldsymbol{b}_i \end{pmatrix} \sim N \left[\begin{pmatrix} \boldsymbol{X}_i \boldsymbol{\beta} \\ 0 \end{pmatrix}, \begin{pmatrix} V_i & Z_i D \\ D Z_i' & D \end{pmatrix} \right].$$

This leads to the conditional distribution,

$$\boldsymbol{b}_i | \boldsymbol{Y}_i \sim N \left(D Z_i' V_i^{-1} \left(\boldsymbol{Y}_i - \boldsymbol{X}_i \boldsymbol{\beta} \right), \Lambda_i \right),$$

where

$$\Lambda_i = D - D Z_i' V_i^{-1} Z_i D = \left(Z_i' \Sigma_i^{-1} Z_i + D^{-1} \right)^{-1}.$$

So the empirical Bayes estimator for $\boldsymbol{b}_i$ is

$$\widetilde{\boldsymbol{b}}_i = E\left[\boldsymbol{b}_i|\boldsymbol{Y}_i, \widehat{\boldsymbol{\alpha}}, \widehat{\boldsymbol{\beta}}\right] = D\left(\widehat{\boldsymbol{\alpha}}\right) \boldsymbol{Z}_i' V_i^{-1}\left(\widehat{\boldsymbol{\alpha}}\right)\left(\boldsymbol{Y}_i - \boldsymbol{X}_i\widehat{\boldsymbol{\beta}}\right).$$

This is also referred to as the *best linear unbiased predictor* for $\boldsymbol{b}_i$ (Robinson, 1991).

5.3.2 Generalized Estimating Equations

5.3.2.1 Model and Inference In cross-sectional studies, if we suppose that the scalar response Y_i comes from the exponential family, then the score equation has the following form:

$$U(\boldsymbol{\beta}) = \sum_{i=1}^{N} \frac{\partial \mu_i}{\partial \boldsymbol{\beta}} v_i^{-1}(Y_i - \mu_i) = 0, \tag{5.9}$$

where $v_i = \text{var}(Y_i)$. With longitudinal data, the idea of generalized estimating equations (GEE) comes from analogy of the score equation (5.9):

$$U(\boldsymbol{\beta}) = \sum_{i=1}^{N} D_i' V_i^{-1}(\boldsymbol{Y}_i - \boldsymbol{\mu}_i) = 0, \tag{5.10}$$

where

$$D_i = \frac{\partial \boldsymbol{\mu}_i}{\partial \boldsymbol{\beta}} = \left(\frac{\partial \mu_{ij}}{\beta_k}\right)_{n_i \times p}$$

and V_i is a working covariance matrix for $\boldsymbol{Y}_i$. We usually decompose the working covariance as $V_i = S_i^{1/2} R_i(\boldsymbol{\alpha}) S_i^{1/2}$, where $S_i = \text{diag}(\text{var}(Y_{i1}), \ldots, \text{var}(Y_{in_i}))$ is a diagonal matrix and $R_i(\boldsymbol{\alpha})$ is a working correlation matrix with a vector of unknown parameters $\boldsymbol{\alpha}$. For the exponential family, S_i follows from the mean–variance relationship, and we only need to specify a plausible guess of $R_i(\boldsymbol{\alpha})$. Common choices of R_i are listed in Table 5.3.

When the sample size N is large, $\widehat{\boldsymbol{\beta}}$, defined as the solution to (5.10), approximately follows the normal distribution with mean $\boldsymbol{\beta}$ and variance $I_0^{-1} I_1 I_0^{-1}$, where

$$I_0 = \sum_{i=1}^{N} D_i' V_i^{-1} D_i$$

TABLE 5.3 Common Choices of Working Correlation R_i with the Cluster Size $n_i = 4$

Independence	AR-1
$\begin{pmatrix} 1 & 0 & 0 & 0 \\ 0 & 1 & 0 & 0 \\ 0 & 0 & 1 & 0 \\ 0 & 0 & 0 & 1 \end{pmatrix}$	$\begin{pmatrix} 1 & \alpha & \alpha^2 & \alpha^3 \\ \alpha & 1 & \alpha & \alpha^2 \\ \alpha^2 & \alpha & 1 & \alpha \\ \alpha^3 & \alpha^2 & \alpha & 1 \end{pmatrix}$
Exchangeable	**Unstructured**
$\begin{pmatrix} 1 & \alpha & \alpha & \alpha \\ \alpha & 1 & \alpha & \alpha \\ \alpha & \alpha & 1 & \alpha \\ \alpha & \alpha & \alpha & 1 \end{pmatrix}$	$\begin{pmatrix} 1 & \alpha_{12} & \alpha_{13} & \alpha_{14} \\ \alpha_{12} & 1 & \alpha_{23} & \alpha_{24} \\ \alpha_{13} & \alpha_{23} & 1 & \alpha_{34} \\ \alpha_{14} & \alpha_{24} & \alpha_{34} & 1 \end{pmatrix}$

and

$$I_1 = \sum_{i=1}^{N} D_i' V_i^{-1} \operatorname{var}(\boldsymbol{Y}_i) V_i^{-1} D_i.$$

In practice, we could replace $\operatorname{var}(\boldsymbol{Y}_i)$ by $(\boldsymbol{Y}_i - \widehat{\boldsymbol{\mu}}_i)(\boldsymbol{Y}_i - \widehat{\boldsymbol{\mu}}_i)'$. Although this is a rather crude estimator for every single $\operatorname{var}(\boldsymbol{Y}_i)$, it leads to a good estimate of I_1 with many replications. The variance $I_0^{-1} I_1 I_0^{-1}$ is also referred to as the sandwich variance estimator. One advantage of GEE is that the inference is always valid even if the working correlation matrix is incorrect. In other words, the only assumption of GEE is the mean structure:

$$E(\boldsymbol{Y}_i | \boldsymbol{X}_i) = \boldsymbol{\mu}_i = g^{-1}(\boldsymbol{X}_i \boldsymbol{\beta}),$$

with the link function g, and the working correlation is not so important. With the correct working correlation, GEE is similar to the likelihood inference, and the sandwich variance simplifies to $I_0^{-1} I_0 I_0^{-1} = I_0^{-1}$, which can be seen as the inverse of the information matrix. The vector of parameters, $\boldsymbol{\beta}$, is interpreted as the population-average effect of the covariates.

5.3.2.2 Estimation of Working Correlation The working correlation may be specified up to the unknown vector of nuisance parameters, $\boldsymbol{\alpha}$, which needs to be estimated from the data. Liang and Zeger (1986) proposed to use a moment estimator for $\boldsymbol{\alpha}$. Their estimator is based on the Pearson residual:

$$r_{ij} = \frac{Y_{ij} - \mu_{ij}(\widehat{\boldsymbol{\beta}})}{\widehat{V}_{ij}^{1/2}},$$

where $\hat{V}_{ij}$ is the estimated variance for Y_{ij}. For example, under the exchangeable working correlation, the nuisance parameter vector $\boldsymbol{\alpha}$ becomes a scalar of within-cluster correlation coefficient α. The moment estimator for α is given by

$$\hat{\alpha} = \frac{1}{N} \sum_{i=1}^{N} \frac{1}{n_i \left(n_i - 1\right)} \sum_{j \neq k} r_{ij} r_{ik}.$$

The standard GEE procedure implemented in most software packages is as follows:

1. Compute the initial estimate of $\widehat{\boldsymbol{\beta}}$ using generalized linear models.
2. Compute the Pearson residual r_{ij} and estimate for $\widehat{\boldsymbol{\alpha}}$ (and thus $\boldsymbol{R}_i$ and V_i).
3. Update the estimate for $\widehat{\boldsymbol{\beta}}$ using the estimated correlation from step 2.
4. Iterate steps 2 and 3 until convergence.

More advanced estimators for $\boldsymbol{\alpha}$ are also proposed in the literature. Interested readers should refer to Prentice (1988), Lipsitz et al. (1991), Carey et al. (1993), and Prentice and Zhao (1991). These approaches generally need to specify another set of estimating equations for the second moments, in order to estimate $\boldsymbol{\alpha}$.

5.3.3 Generalized Linear Mixed Models

5.3.3.1 Model Setting In Section 5.3.1, we introduced the mixed model for linear regression. The same idea could be applied to generalized linear models, when multiple observations on each subject are taken over time. Suppose that given the random effects $\boldsymbol{b}_i$, all the responses Y_{ij} are independent and follow a distribution in the exponential family. The generalized linear mixed model assumes that

$$\begin{aligned}
&Y_{ij} | \boldsymbol{X}_{ij}, \boldsymbol{b}_i \sim \text{distribution}, \\
&Y_{ij} \perp Y_{ik} | \boldsymbol{b}_i \text{: independence}, \\
&g\left\{E\left(Y_{ij} | \boldsymbol{X}_{ij}, \boldsymbol{b}_i\right)\right\} = \boldsymbol{X}'_{ij} \beta + \boldsymbol{Z}'_{ij} \boldsymbol{b}_i, \\
&\boldsymbol{b}_i \sim N(\mathbf{0}, D),
\end{aligned}$$

where "$\perp$" denotes independence. Note that the vector of parameters, $\boldsymbol{\beta}$, is interpreted as the *subject-specific* effect of the covariates (conditional on the random effects). In contrast, the parameters in GEE have a completely different marginal interpretation—the *population-average* effect of the covariates. Generally, GEE and GLMM are not compatible models: if one integrates out $\boldsymbol{b}_i$ in $E\left(Y_{ij} | \mathbf{X}_{ij}, \boldsymbol{b}_i\right)$ according to the GLMM specification, then the marginal mean $E\left(Y_{ij} | \mathbf{X}_{ij}\right)$ usually does not possess the generalized linear form.

Let $f_{ij}\left(y_{ij}|\boldsymbol{b}_i;\boldsymbol{\beta}\right)$ be the conditional density of Y_{ij} given $\boldsymbol{b}_i$, and $f\left(\boldsymbol{b}_i; D\right)$ be the marginal density of $\boldsymbol{b}_i$. Then, the likelihood function of $\boldsymbol{\beta}$ and D is

$$L(\boldsymbol{\beta}, D) = \prod_{i=1}^{N} \int \left(\prod_{j=1}^{n_i} f_{ij}\left(y_{ij}|\boldsymbol{b}_i;\boldsymbol{\beta}\right) \right) f\left(\boldsymbol{b}_i; D\right) \mathrm{d}\boldsymbol{b}_i. \tag{5.11}$$

If f_{ij} is the normal density, the integral could be calculated analytically, which is the LMM in Section 5.3.1. Otherwise, approximations are needed to evaluate the likelihood function.

5.3.3.2 Approximation Methods We introduce four methods to approximate the integral in (5.11), namely, penalized quasi-likelihood (PQL), marginal quasi-likelihood (MQL), Laplace approximation, and Gaussian quadrature. The technical details of the four approximation methods are shown in Appendix 5.A.

Both the PQL and MQL use the idea of linearization in the generalized linear models (McCullagh and Nelder, 1989). Their computation is relatively simple. However, the approximation is rather crude with a small cluster size, that is, a small number of observations per subject. The performance of both methods is particularly poor for the binary outcome, usually yielding serious underestimation of the fixed effects and variance components. The MQL approximation adopts a Taylor series expansion around $\boldsymbol{b}_i = 0$, so it is recommended only when the variance of the random effects is very small. Higher order PQL and MQL approaches could potentially reduce the bias (Rodriguez and Goldman, 1995; Goldstein and Rasbash, 1996), but they are currently not implemented in R, STATA, or SAS.

The Laplace approximation uses a multivariate normal density to approximate the integrand for each subject. The Laplace approximation is also quite fast for computation, but has similar limitations to the PQL method. Since the approximation is performed on each subject, it would be more accurate with more observations per subject. With binary data, the approximation often behaves poorly. Raudenbush et al. (2000) developed higher order Laplace approximations, which have a good improvement on the accuracy and are computationally fast.

The Gaussian quadrature method is the slowest among all the approximation methods, but it leads to the best approximation of the likelihood and hence more reliable estimates of the GLMM. As long as the computation facility allows, the Gaussian quadrature is always recommended for practical use.

The first-order PQL and MQL, the first-order Laplace approximation, and Gaussian quadrature methods have all been implemented in standard software packages, for example, R, STATA, and SAS. Other statistical packages such as MLwiN (Rasbash et al., 2012) and HLM (Raudenbush et al., 2000) are also available for some higher order approaches. Breslow (2003) gave a comprehensive review of the existing methods for estimating GLMM.

5.3.4 Time-Dependent Covariates

In the discussion of the methods for longitudinal data analysis, so far we are interested in the mean of Y_{ij} given X_{ij}, which is also called the cross-sectional mean. In practice, however, it is often the case that the mean of Y_{ij} can be affected by previous observations of covariates, $X_{i1}, \ldots, X_{i,j-1}$. For example, when studying the association between lung function and air pollution, there is usually a lag between the peak of the pollutant and the change of lung function, so the quantity of interest might be $E(Y_{ij}|X_{i,j-1})$. When studying the effect of a new drug in clinical trials, it is more reasonable to assume a cumulative treatment effect, so the interest is in $E(Y_{ij}|X_{i1}, \ldots, X_{ij})$, where we condition on the whole treatment history.

We introduce an important concept for the time-dependent covariates—*exogeneity*. A covariate process is *exogenous* to the outcome process if the distribution of current covariate value is completely determined by its previous values, that is,

$$P(X_{ij}|H_i^Y(j), H_i^X(j-1), W_i) = P(X_{ij}|H_i^X(j-1), W_i),$$

where $H_i^Y(j) = \{Y_{i1}, \ldots, Y_{ij}\}$ and $H_i^X(j) = \{X_{i1}, \ldots, X_{ij}\}$ are the past observations of Y and X until time j and W_i is the vector of baseline or time-invariant covariates. If the conditional independence is not satisfied, then the covariate process is said to be *endogenous*. For example, in a clinical trial with a crossover design, if the treatment is prescribed at the start of the trial, then the treatment variable is exogenous. On the contrary, if the treatment assignment changes according to the patients' response to the drug, then the treatment variable is endogenous. In general, one can check for endogeneity by regressing X_{ij} on both $H_i^Y(j)$ and $H_i^X(j-1)$.

The advantage of exogeneity is that the joint likelihood of $\boldsymbol{Y}$ and $\boldsymbol{X}$ can be factored out:

$$\begin{aligned} f(\boldsymbol{X}_i, \boldsymbol{Y}_i|W_i) &= \left\{ \prod_{j=1}^{n_i} f(Y_{ij}|H_i^Y(j-1), H_i^X(j), W_i) \right\} \\ &\quad \times \left\{ \prod_{j=1}^{n_i} f(X_{ij}|H_i^X(j-1), W_i) \right\} \\ &= f_{Y|X} \times f_X, \end{aligned}$$

where n_i is the number of observations for subject i, $\boldsymbol{X}_i = (X_{i1}, \ldots, X_{in_i})'$, and $\boldsymbol{Y}_i = (Y_{i1}, \ldots, Y_{in_i})'$. Therefore, if one is interested only in $f_{Y|X}$, no assumptions on the covariate process are needed. Exogeneity also suggests that

$$E\left(Y_{ij}|X_{i1}, \ldots, X_{in_i}, W_i\right) = E\left(Y_{ij}|X_{i1}, \ldots, X_{ij}, W_i\right).$$

When a covariate is endogenous, we need to be careful in choosing the meaningful model of interest.

Pepe and Anderson (1994) have shown that when including endogenous variables into the regression model, GEE could be biased unless *either* of the following assumptions holds:

1. Full covariate conditional mean assumption:

$$E\left(Y_{ij}|\boldsymbol{X}_i\right) = E\left(Y_{ij}|X_{ij}\right).$$

2. A working-independent correlation assumption.

5.4 MISSING DATA SETTINGS AND SIMPLE METHODS

5.4.1 Setup

In Section 5.3, we summarized commonly used methods for analyzing longitudinal data without missingness. In this section, we will introduce statistical methods to handle longitudinal missing data. Sections 5.5–5.7 will focus on missing outcome only. The scenario of missing both outcome and covariates is discussed in Sections 5.8–5.10.

The pattern of the missing data can be classified as monotone and nonmonotone missingness. As a special case of monotone missingness, *dropout* means that if Y_{ij} is missing, all the observations after time j are missing. Otherwise, the missingness is said to be *intermittent*.

In Chapter 1, we distinguished the missingness mechanisms as missing completely at random (MCAR), missing at random (MAR), and missing not at random (MNAR). For longitudinal missing data, we further define *covariate-dependent missingness* (CDM): an MAR missingness process is CDM if the probability of nonresponse depends only on covariate X and not on Y. Chapter 5 is entirely based on the MAR assumption, and MNAR models will be introduced in Chapter 7.

5.4.2 Simple Methods

One approach for handling missing longitudinal data is simply to remove all the subjects with any missing observations, known as the *complete-case analysis*. Although easily implemented, this method is usually biased when the missingness mechanism is MAR or MNAR. Even under the MCAR situation, deleting subjects with incomplete observations could lead to substantial loss of information, and hence result in inefficient estimators.

Another way to handle the missing data is single imputation, that is, to replace the missing value with some kind of "guess." One of the most widely used single imputation methods is *last observation carried forward* (LOCF). This method replaces every missing value by the last observed value of the same subject. The LOCF method makes a very strong assumption that the value of the outcome remains unchanged after missing. When the missing data are due to cure, this assumption might be appropriate. But in many other settings, LOCF is likely to yield biased results. Saha and Jones (2009) derived an explicit expression for the bias of LOCF under a linear mixed effects model. They found that the bias for the two-group mean comparison is more severe if the time to drop out is shorter, or the dropout probability is more strongly associated with the individual random effects. They also noticed that the variance is always underestimated. Other single imputation methods, such as mean imputation or hot deck imputation, could be used as well. The disadvantage of all these single imputation methods is that the imputed observations are treated as known, so the inference does not reflect the sample variability without special adjustments.

5.5 LIKELIHOOD APPROACH

Let R_{ij} be the missingness indicator of Y_{ij}, which takes the value of 1 if Y_{ij} is observed and 0 otherwise. We write $\boldsymbol{R}_i \equiv (R_{i1}, \ldots, R_{in_i})'$. The outcome vector $\boldsymbol{Y}_i$ could be partitioned into $(\boldsymbol{Y}_i^{\text{obs}}, \boldsymbol{Y}_i^{\text{mis}})$. For the ith subject, its contribution to the likelihood is

$$\begin{aligned}
&\int f(\boldsymbol{Y}_i, \boldsymbol{R}_i | \boldsymbol{X}_i; \boldsymbol{\theta}, \boldsymbol{\eta}) \mathrm{d}\boldsymbol{Y}_i^{\text{mis}} \\
&= \int f(\boldsymbol{Y}_i | \boldsymbol{X}_i; \boldsymbol{\theta}) f(\boldsymbol{R}_i | \boldsymbol{Y}_i^{\text{obs}}, \boldsymbol{Y}_i^{\text{mis}}, \boldsymbol{X}_i; \boldsymbol{\eta}) \mathrm{d}\boldsymbol{Y}_i^{\text{mis}} \\
&= \int f(\boldsymbol{Y}_i | \boldsymbol{X}_i; \boldsymbol{\theta}) f(\boldsymbol{R}_i | \boldsymbol{Y}_i^{\text{obs}}, \boldsymbol{X}_i; \boldsymbol{\eta}) \mathrm{d}\boldsymbol{Y}_i^{\text{mis}} \\
&= f(\boldsymbol{R}_i | \boldsymbol{Y}_i^{\text{obs}}, \boldsymbol{X}_i; \boldsymbol{\eta}) \int f(\boldsymbol{Y}_i | \boldsymbol{X}_i; \boldsymbol{\theta}) \mathrm{d}\boldsymbol{Y}_i^{\text{mis}} \\
&= f(\boldsymbol{R}_i | \boldsymbol{Y}_i^{\text{obs}}, \boldsymbol{X}_i; \boldsymbol{\eta}) f(\boldsymbol{Y}_i^{\text{obs}} | \boldsymbol{X}_i; \boldsymbol{\theta}).
\end{aligned}$$

The second equality above holds because of the MAR assumption. Here, $\boldsymbol{\theta}$ is the vector of parameters of interest and $\boldsymbol{\eta}$ is the vector of nuisance parameters involved in modeling the missingness mechanism. The last equality suggests that $\boldsymbol{\eta}$ and $\boldsymbol{\theta}$ are separated in the likelihood, so the inference for $\boldsymbol{\theta}$ can be simply based on $f(\boldsymbol{Y}_i^{\text{obs}} | \boldsymbol{X}_i; \boldsymbol{\theta})$. In other words, the *available case analysis* with the full likelihood approach would produce consistent and efficient estimators. The LMM and GLMM are both likelihood methods; therefore, when the missing data are present, we could fit the same model on $\boldsymbol{Y}_i^{\text{obs}}$ without any special treatment.

However, the available case analysis does not always work if the analysis method is not likelihood based. Typically, a GEE model on $\boldsymbol{Y}_i^{\text{obs}}$ is unbiased only when the missingness mechanism is CDM. A heuristic proof is provided as follows. The estimating equation for the available case analysis is

$$\sum_{i=1}^{N} D_i' V_i^{-1} \Delta_i (\mathbf{Y}_i - \boldsymbol{\mu}_i) = 0, \tag{5.12}$$

where

$$\Delta_i = \text{diag}\{R_{i1}, \ldots, R_{in_i}\}.$$

We have

$$\begin{aligned}
&E_{\boldsymbol{R}_i,\boldsymbol{Y}_i|\mathbf{X}_i}\left[D_i'V_i^{-1}\Delta_i(\boldsymbol{Y}_i - \boldsymbol{\mu}_i)\middle|\boldsymbol{X}_i\right] \\
&= E_{\boldsymbol{Y}_i|\mathbf{X}_i}\left\{E_{\boldsymbol{R}_i|\boldsymbol{Y}_i,\boldsymbol{X}_i}\left[D_i'V_i^{-1}\Delta_i(\boldsymbol{Y}_i - \boldsymbol{\mu}_i)\right]\right\} \\
&= \begin{cases} E_{\boldsymbol{Y}_i|\mathbf{X}_i}\left[D_i'V_i^{-1}\Pi_i(\boldsymbol{X}_i, \boldsymbol{Y}_i^{\text{obs}})(\boldsymbol{Y}_i - \boldsymbol{\mu}_i)\right], & \text{MAR}, \\ E_{\boldsymbol{Y}_i|\mathbf{X}_i}\left[D_i'V_i^{-1}\Pi_i(\boldsymbol{X}_i)(\boldsymbol{Y}_i - \boldsymbol{\mu}_i)\right], & \text{CDM}, \end{cases} \\
&= \begin{cases} E_{\boldsymbol{Y}_i|\mathbf{X}_i}\left[D_i'V_i^{-1}\Pi_i(\boldsymbol{X}_i, \boldsymbol{Y}_i^{\text{obs}})(\boldsymbol{Y}_i - \boldsymbol{\mu}_i)\right], & \text{MAR}, \\ 0, & \text{CDM}, \end{cases}
\end{aligned}$$

where $\Pi_i(\cdot)$ is the expectation of Δ_i given $\boldsymbol{Y}_i$ and $\boldsymbol{X}_i$. Only under CDM, $\Pi_i(\cdot)$ does not depend on $\boldsymbol{Y}_i^{\text{obs}}$, and thus the expectation of the estimation equation is 0. Under MAR, the expectation would depend on the second moment of $\boldsymbol{Y}_i^{\text{obs}}$ and is not necessarily 0. In the situations where $\boldsymbol{R}_i$ depends on both $\boldsymbol{Y}_i^{\text{obs}}$ and $\boldsymbol{X}_i$, a weighted GEE approach could be used, as shown in Section 5.6.

5.5.1 Example: IMPACT Study

Table 5.4 lists percentages of missingness by treatment and month. It appears that the intervention arm had a slightly lower percentage of missing data in SCL-20 scores, but the differences did not appear to be significant between the two arms.

Note that we model the IMPACT data by (5.1). We carry out some inferences in the following sections. There are more efficient imputation techniques available for monotone missingness. To illustrate the method, we first drop the subjects with intermittent missing outcome values and then return to the full sample toward the end.

5.5.1.1 Complete-Case Analysis We first consider the subjects with complete data ($N = 1269$). The intervention arm is estimated to have 0.3115 more reduction in change from baseline for the SCL-20 score than the usual care group.

TABLE 5.4 Percentages of Missingness by Treatment and Month

	Usual Care	Intervention	Both Arms
Month 3	12.4	10.26	11.33
Month 6	14.08	11.59	12.83
Month 12	18.44	15.01	16.71
Month 18	22.12	19.43	20.77
Month 24	23.46	22.08	22.77

TABLE 5.5 Estimates Based on the Complete-Case Analysis

Parameter	Estimate	Robust SE	95% CI
β_1	1.689	0.025	(1.639, 1.738)
β_2	−0.347	0.024	(−0.395, −0.299)
β_3	−0.006	0.034	(−0.073, 0.060)
β_4	−0.312	0.034	(−0.378, −0.245)
$\sqrt{g_{11}} = \text{SD}(b_{1i})$	0.452	0.016	(0.422, 0.484)
$\sqrt{g_{22}} = \text{SD}(b_{2i})$	0.412	0.019	(0.377, 0.450)
$\text{Corr}(b_{1i}, b_{2i})$	−0.235	0.051	(−0.332, −0.134)

For the complete-case analysis, the estimated fixed effects and variance components from the linear mixed effects model are displayed in Table 5.5. A test of the null hypothesis, $H_0 : \beta_4 = 0$, indicates that there is a significant group × period interaction at the 0.05 level. These results suggest that there are differences between the two treatment arms in terms of subject-specific changes in the expected SCL-20 scores. In particular, there is a greater reduction in the expected SCL-20 score from baseline for patients treated with IMPACT care (compared with patients who received the usual care).

The estimated variance components for the random intercept and slope indicate that there is substantial variability in the baseline SCL-20 score in the study population and also substantial variability in the patient-to-patient changes in the SCL-20 score in response to the treatment. For example, the estimated standard deviation of the random intercepts, $\sqrt{\hat{g}_{11}} = 0.452$, implies that there is substantial patient-to-patient variability in terms of their baseline SCL-20 score, since ~95% of the patients have a baseline SCL-20 score that varies from $1.689 - 1.96 \times 0.452$ to $1.689 + 1.96 \times 0.452$, or from 0.803 to 2.575. Similarly, there is moderate heterogeneity in the patient-to-patient changes in the SCL-20 score. Finally, the negative correlation among the random intercepts and slopes is significant, indicating that the expected change in the SCL-20 score is directly related to the baseline SCL-20 score, and patients with higher baseline SCL-20 scores tend to have smaller changes in SCL-20 scores, compared with those with lower baseline SCL-20 scores.

5.5.1.2 Available Data Analysis One advantage with the mixed effects model is that it allows all available data to be included in the analysis. In the following example, we again consider the sample with monotone missingness ($N = 1639$), and the results based on the available data analysis are reported in Table 5.6. Compared to the complete-case analysis, the estimated intervention effect (*group* × *period*) is smaller with smaller standard error too.

5.5.1.3 Last Observation Carried Forward Analysis with Monotone Missingness A popular and less sophisticated approach is the last observation carried forward approach. By its name, the last observed value is used to replace all

TABLE 5.6 Estimates Based on the Available Data Analysis

Parameter	Estimate	Robust SE	95% CI
β_1	1.686	0.022	(1.644, 1.729)
β_2	−0.335	0.022	(−0.378, −0.291)
β_3	−0.003	0.030	(−0.056, 0.062)
β_4	−0.295	0.031	(−0.356, −0.235)
$\sqrt{g_{11}} = \mathrm{SD}(b_{1i})$	0.455	0.014	(0.428, 0.484)
$\sqrt{g_{22}} = \mathrm{SD}(b_{2i})$	0.414	0.018	(0.380, 0.451)
$\mathrm{Corr}(b_{1i}, b_{2i})$	−0.200	0.048	(−0.292, −0.105)

TABLE 5.7 Estimates Based on the LOCF Analysis

Parameter	Estimate	Robust SE	95% CI
β_1	1.686	0.022	(1.644, 1.729)
β_2	−0.313	0.021	(−0.354, −0.271)
β_3	−0.003	0.030	(−0.056, 0.062)
β_4	−0.264	0.030	(−0.323, −0.206)
$\sqrt{g_{11}} = \mathrm{SD}(b_{1i})$	0.481	0.013	(0.458, 0.508)
$\sqrt{g_{22}} = \mathrm{SD}(b_{2i})$	0.449	0.016	(0.419, 0.482)
$\mathrm{Corr}(b_{1i}, b_{2i})$	−0.193	0.040	(−0.270, −0.113)

missing values if a subject drops out. The results of the LOCF analysis are listed in Table 5.7. Note that the estimated intervention effect is much smaller than both complete-case analysis and available data analysis, suggesting that such an approach is very prone to bias.

5.6 INVERSE PROBABILITY WEIGHTED GEE WITH MAR DROPOUT

The standard GEE model is valid only under the CDM assumption. Under a weaker assumption of MAR, two different formulations of the inverse probability weighted GEE were proposed by Robins et al. (1995) and Fitzmaurice et al. (1995), which we refer to as the IPWGEE1 and IPWGEE2 methods, respectively. The difference is that IPWGEE1 (Robins et al., 1995) uses observation-level weights and IPWGEE2 (Fitzmaurice et al., 1995) uses cluster-level weights.

5.6.1 Modeling the Selection Probability

The dropout pattern implies that if $R_{ij} = 0$, then $R_{i,j+1} = 0$. Assume that $R_{i1} = 1$ for the first time point. Let $M_i = 1 + \sum_{j=1}^{n_i} R_{ij}$, which indicates the time of dropout and takes the value between 2 and $n_i + 1$. The maximum value $M_i = n_i + 1$ corresponds to a subject with complete longitudinal observations.

The *continuation ratio model* (CRM) can be used to estimate the selection model. Define

$$\lambda_{ij} \equiv \Pr(R_{ij} = 1 \big| R_{i,j-1} = 1, \boldsymbol{X}_i, Y_{i1,} \ldots, Y_{i,j-1}).$$

Assume that

$$\text{logit}\left(\lambda_{ij}\right) = \boldsymbol{W}_{ij}'\boldsymbol{\alpha},$$

where $\boldsymbol{W}_{ij}$ is the design matrix expanded by $\boldsymbol{X}_i, Y_{i1,} \ldots, Y_{i,j-1}$. The CRM can be regarded as a discrete-time survival model, which estimates the dropout probability at the current time point, given the dropout not occurring at previous time points. Note that

$$\begin{aligned} \upsilon_{ij} &\equiv \Pr(M_i = j | \boldsymbol{X}_i, \boldsymbol{Y}_i^{\text{obs}})] \\ &= \begin{cases} (1 - \lambda_{ij}) \times \prod_{k=2}^{j-1} \lambda_{ik}, & \text{if } j \leq n_i, \\ \prod_{k=2}^{n_i} \lambda_{ik}, & \text{if } j = n_i + 1, \end{cases} \end{aligned}$$

which is the likelihood contribution for the ith subject in estimating the selection model. The MLE for the continuation ratio model are readily available from many standard statistical packages.

5.6.2 IPWGEE1 and IPWGEE2

Robins et al. (1995) proposed to weight the observation Y_{ij} using

$$\frac{R_{ij}}{\Pr(R_{ij} = 1 \big| \boldsymbol{X}_i, \boldsymbol{Y}_i^{\text{obs}})},$$

in which the denominator is denoted as π_{ij}. Note that

$$\begin{aligned} \pi_{ij} &= \Pr(R_{i1} = \cdots = R_{ij} = 1 \big| \boldsymbol{X}_i, \boldsymbol{Y}_i^{\text{obs}}) \\ &= \prod_{k=2}^{j} \lambda_{ik}. \end{aligned}$$

The second equation is due to factoring out the joint distribution of $R_{i1}, \ldots, R_{ij}$. This leads to the following estimating equation for IPWGEE1:

$$\sum_{i=1}^{N} D_i' V_i^{-1} \Delta_i^* (\mathbf{Y}_i - \boldsymbol{\mu}_i) = 0, \tag{5.13}$$

where

$$\Delta_i^* = \text{diag}\left\{ \frac{R_{i1}}{\hat{\pi}_{i1}}, \cdots, \frac{R_{in_i}}{\hat{\pi}_{in_i}} \right\}.$$

Here, $\hat{\pi}_{ij}$ is the estimated selection probability from the continuation ratio model. Because λ_{ik} is always smaller than 1, π_{ij} decreases as j increases. In other words, more weights are assigned to the later observations.

An alternative approach is to weight each subject by the inverse of the probability of dropping out at the observed time (Fitzmaurice et al., 1995). In this formulation, the estimating equation for IPWGEE2 becomes

$$\sum_{i=1}^{N} D_i' V_i^{-1} \Delta_i^{**} (\boldsymbol{Y}_i - \boldsymbol{\mu}_i) = 0, \tag{5.14}$$

where

$$\Delta_i^{**} = \text{diag}\left\{ \frac{R_{i1}}{\hat{\upsilon}_{im_i}}, \cdots, \frac{R_{in_i}}{\hat{\upsilon}_{im_i}} \right\}.$$

Here, $\hat{\upsilon}_{im_i}$ is the estimated dropout probability from the continuation ratio model, with m_i being the observed dropout time for subject i.

If the true value of λ_{ik} is used instead of being estimated from the continuation ratio model, the standard theory of estimating equations leads to the sandwich variance estimator for $\widehat{\boldsymbol{\theta}}$:

$$\text{var}(\widehat{\boldsymbol{\theta}}) \approx I_N^{-1} \left(\sum_{i=1}^{N} U_i U_i' \right) I_N^{-1}. \tag{5.15}$$

Here, U_i is the summand in (5.13) or (5.14) and I_N is the derivative of $\sum_i U_i$:

$$U_i = \begin{cases} D_i' V_i^{-1} \Delta_i^* (\boldsymbol{Y}_i - \boldsymbol{\mu}_i), & \text{for IPWGEE1,} \\ D_i' V_i^{-1} \Delta_i^{**} (\boldsymbol{Y}_i - \boldsymbol{\mu}_i), & \text{for IPWGEE2,} \end{cases}$$

$$I_N = \sum_i \frac{\partial}{\partial \boldsymbol{\theta}} U_i.$$

However, estimating the selection model contributes an additional term to the sandwich variance formula (5.15), and a rigorous proof is given in Robins et al. (1995). For simplicity, we just ignore the additional term, and the sandwich variance formula seems to work well in our simulation study.

5.6.3 A Simulation Study

We conduct a simulation study to compare the two IPWGEE methods with the naive GEE estimator that ignore the missing data. Consider $N = 400$ subjects, each with $n_i = 5$ observations. The true model is generated by a mixed effects model:

$$Y_{ij}^* = 4 + 2 \times \text{Age}_i + 1.2 \times \text{Time}_{ij} + 1.5 \times \text{Treatment}_i + b_i + \varepsilon_{ij},$$
$$Y_{ij} = 1 \text{ if } Y_{ij}^* > 10, \text{ and 0 otherwise.}$$

The Age variable follows the uniform distribution between 0 and 1, indicating the baseline age of the subject; the Time variable takes values from 1 up to 5, being the time of observation; the Treatment variable is equal to 0 for half of the subjects and 1 for the other half; b_i is the random intercept and ε_{ij} is the random error, both of which independently follow a normal distribution with variance 1.2^2 and 1, respectively. The marginal mean model is a probit model:

$$\Phi^{-1}(\Pr(Y_{ij} = 1|X_{ij}))$$
$$= -3.841 + 1.280 \times \text{Age}_i + 0.768 \times \text{Time}_{ij} + 0.960 \times \text{Treatment}_i,$$

where Φ is the cumulative distribution function of a standard normal distribution. The dropout process is generated from a continuation ratio model:

$$\text{logit}(R_{ij} = 1|R_{i,j-1} = 1, X_{ij}, Y_{i,j-1})$$
$$= 2.5 \times I(j = 2) + 3 \times I(j = 3) + 2 \times I(j = 4) + 2.2 \times I(j = 5)$$
$$- 1.3 \times \text{Age}_i + 0.8 \times \text{Treatment}_i - 2 \times Y_{i,j-1}.$$

The averaged parameter estimation and the 95% CI coverage are shown in Table 5.8. Both IPWGEE1 and IPWGEE2 are consistent and the CI coverage rates are close to the nominal level. The ordinary GEE is biased as we expected.

TABLE 5.8 The Average of Estimated Regression Parameters (Coverage Rate in Percentage of a CI with Nominal Level 95%) for the Full Data Estimator, Available Case Analysis (GEE), IPWGEE1, and IPWGEE2

	True	Full Data	GEE	IPWGEE1	IPWGEE2
Intercept	−3.841	−3.867 (94.6)	−3.697 (84.6)	−3.871 (92.6)	−3.892 (93.8)
Age	1.280	1.282 (95.0)	1.157 (89.0)	1.277 (93.2)	1.261 (95.0)
Time	0.768	0.774 (94.4)	0.716 (71.6)	0.775 (94.0)	0.781 (91.8)
Treatment	0.960	0.969 (96.2)	0.949 (94.0)	0.973 (95.4)	0.991 (94.0)

TABLE 5.9 Estimates Based on the IPWGEE1 Analysis

Parameter	Estimate	Robust SE	95% CI
β_1	1.686	0.022	(1.644, 1.729)
β_2	−0.327	0.023	(−0.372, −0.282)
β_3	−0.003	0.030	(−0.056, 0.062)
β_4	−0.314	0.032	(−0.377, −0.251)

More extensive simulations that compare IPWGEE1 and IPWGEE2 can be found in Preisser et al. (2002). They found that IPWGEE2 is considerably less efficient than IPWGEE1, while both are consistent. When the dropout probability is low, IPWGEE2 could be quite inefficient and anticonservative (larger type 1 error rate than expected) for small samples.

5.6.4 Example: IMPACT Study

We next illustrate how to implement the two IPWGEE methods to analyze longitudinal data with dropouts (therefore, the missing data pattern is monotone).

With the Stata code presented in Appendix 5.A, we first estimate the probability that an outcome is observed at a specific time point, given the prior observed outcome variables and some selected baseline characteristics using the continuation ratio model. We then generate both observation-level weights and cluster (subject)-level weights based on the estimated probabilities. It is worth noting that the uncertainty in the estimated weights will not be accounted for in the IPWGEE analysis. One way to get around this is to apply a bootstrap method.

The results obtained from the two IPWGEE methods are listed in Tables 5.9 and 5.10. It can be seen that the estimated treatment effect ($\beta_4 = -0.3140$) of the IPWGEE1 is close to what we had before. The estimated intervention effect ($\beta_4 = -0.2044$) of the IPWGEE2 is much smaller with a much wider confidence interval, suggesting that the estimator using cluster-level weights may not be stable in this case.

TABLE 5.10 Estimates Based on the IPWGEE2 Analysis

Parameter	Estimate	Robust SE	95% CI
β_1	1.679	0.041	(1.599, 1.759)
β_2	−0.374	0.050	(−0.472, −0.276)
β_3	−0.013	0.060	(−0.105, 0.131)
β_4	−0.204	0.070	(−0.343, −0.066)

5.7 EXTENSION TO NONMONOTONE MISSINGNESS

It is common in longitudinal studies that subjects may miss intermittent visits, or some measurements in a particular visit are missing, for example, item nonresponse. In this case, likelihood-based inferences remain valid, but IPWGEE methods need to be modified. The modified IPWGEE is referred to as IPWGEE3.

We continue to use R_{ij} as the missing indicator, but $R_{ij} = 0$ no longer implies $R_{i,j+1} = 0$. The extension is based on IPWGEE2, where the weight is the inverse of the probability of observing the current missing pattern, that is,

$$\lambda_i \equiv \Pr(R_{i1}, \ldots, R_{in_i} | \boldsymbol{X}_i, \boldsymbol{Y}_i^{\text{obs}}). \tag{5.16}$$

A simpler case is when one is willing to assume conditional independence for R_{ij} within the same subject i. We can use logistic regression to estimate $\Pr(R_{ij} | \boldsymbol{X}_i, \boldsymbol{Y}_i^{\text{obs}})$, and multiplying over all j yields the joint probability (5.16). An example might be in the clustered survey sampling, where i represents a class and j represents a student in the class. Then after conditioning on some student characteristics, the nonresponse of one student is independent of other students.

However, conditional independence is less likely in longitudinal studies. A subject may be more prone to missingness due to some of his/her unobserved characteristics. So, the missing indicators on different occasions are often positively correlated. For example, in the NACC UDS data, a patient with more serious cognitive impairment tends to miss more test results. Of course, we could adjust for the disease severity (e.g., clinical dementia rating sum of boxes) and assume conditional independence anyway. But a more realistic modeling framework would be a random effects model. The joint distribution (5.16) is indeed the ith subject's contribution to the likelihood after the integration over the random effects distribution, which is the by-product of estimating the random effects model using the Gaussian quadrature.

λ_i estimated, we can weight the available case GEE with $1/\lambda_i$ for subject i, which we call IPWGEE3a if independent missingness is assumed and IPWGEE3b if the random effects model is used to estimate the selection probability. The same variance formula (5.15) for IPWGEE1 and IPWGEE2 can be directly applied to IPWGEE3.

5.8 MULTIPLE IMPUTATION

The likelihood approach and the weighted estimating equation approach provide valid inference with the monotone missingness pattern (dropout), and their extensions to the intermittent missingness are straightforward. However, if some of the covariates are subject to missingness too, neither approach has simple extensions. Multiple imputation, as we have seen in Chapter 4, is now a flexible choice to deal with the complicated missing patterns. As long as the imputation model is carefully specified to approximate the true data generation mechanism (although usually hard in practice), the inference immediately simplifies to the analysis model of the full data, and is usually valid using Rubin's combination rule. As commented by van Buuren and Groothuis-Oudshoorn (2011), a good imputation model should reflect the true data generation process, preserve the correlations of the variables, and include uncertainty about the correlations. In the context of longitudinal studies, the difficulty is that the imputation model should take into account the multivariate relationship of the missing variables at one time point, as well as their longitudinal patterns. In this section, we introduce two major ways to set up the imputation model: joint modeling and conditional modeling.

5.8.1 Joint Imputation Model

Schafer (1997a) proposed a Bayesian framework for MI using a multivariate linear mixed model. An important assumption he made was that all the variables subject to missingness follow a multivariate normal distribution. We will outline this method below.

We introduce the following notations that are different from previous sections. Let y_i be an $n_i \times r$ matrix, where each row of y_i corresponds to an observation time point for subject i, and each column corresponds to a variable subject to missingness. So this ''outcome'' matrix includes all the variables to be imputed: not only the outcome variables for the main analysis, but also the covariates with missing values. Assume that y_i follows a multivariate linear mixed model:

$$y_i = X_i \beta + Z_i b_i + \varepsilon_i, \tag{5.17}$$

where X_i is an $n_i \times p$ matrix of covariates that are not subject to missingness, Z_i is an $n_i \times q$ design matrix for the random components, β is a $p \times r$ matrix of the fixed effects, and b_i is a $q \times r$ matrix of the random effects. We further assume that each row of the residual matrix, ε_i, independently follows $N(0, \Sigma)$, while the random effects are distributed as

$$b_i^V \sim N(0, \Psi)$$

independently for $i = 1, \ldots, N$. The superscript V indicates vectorization of a matrix by stacking its columns. By integration over b_i, the marginal model becomes

$$y_i^V \sim N((X_i \beta)^V, (I_r \otimes Z_i)\Psi(I_r \otimes Z_i)^T + \Sigma \otimes I_{n_i}),$$

where I_r is the identity matrix of r dimensions and $\otimes$ is the Kronecker product. We partition y_i into $y_{i(\text{obs})}$ and $y_{i(\text{mis})}$, denoting the observed and missing parts of y_i, respectively. Let $Y_{\text{obs}} = \{y_{i(\text{obs})}\}_{i=1}^N$ and $Y_{\text{mis}} = \{y_{i(\text{mis})}\}_{i=1}^N$. Let $B = (b_1^V, \ldots, b_N^V)'$

and $\theta = (\beta, \Sigma, \Psi)$. The Gibbs sampler updates b_i, θ, and $y_{i(\text{mis})}$ iteratively, in the following three steps:

$$\text{(i) } b_i^{(t+1)} \sim P(b_i \mid Y_{\text{obs}}, Y_{\text{mis}}^{(t)}, \theta^{(t)}), \quad i = 1, \ldots, N.$$

$$\text{(ii) } \theta^{(t+1)} \sim P(\theta \mid Y_{\text{obs}}, Y_{\text{mis}}^{(t)}, B^{(t+1)}).$$

$$\text{(iii) } y_{i(\text{mis})}^{(t+1)} \sim P(y_{i(\text{mis})} \mid Y_{\text{obs}}, B^{(t+1)}, \theta^{(t+1)}), \quad i = 1, \ldots, N.$$

These three steps give the sequences $\{\theta^{(t)}\}$ and $\{Y_{\text{mis}}^{(t)}\}$, which converge in distribution to $P(\theta|Y_{\text{obs}})$ and $P(Y_{\text{mis}}|Y_{\text{obs}})$, respectively. The first step is to draw a sample from a multivariate normal distribution. The second step is a full Bayesian analysis pretending we have the complete data. With conjugate priors on θ, it can be shown that the posterior of β is from a multivariate normal distribution, whereas the variance components Σ^{-1} and Ψ^{-1} are from Wishart distributions. The last step is to draw a sample from the multivariate normal distribution, because of the normality assumption of the multivariate linear mixed model. The whole algorithm is implemented in the `R` library `PAN`.

The Gibbs sampler resembles the idea of data augmentation (Tanner and Wong, 1987) in drawing random sample from the posterior predictive distribution. The major difference is that the estimation of θ is not of high priority, as the multiple imputation model (5.17) itself is a nuisance. The interest here is only in drawing samples from the posterior predictive distribution $P(Y_{\text{mis}}|Y_{\text{obs}})$. In practice, after a suitable number of burn-in cycles, samples from every kth iteration are stored as the imputed data, in order to remove the autocorrelation within the samples. The advantage of the joint imputation model is that proper imputation (in the sense that the imputation comes from the posterior predictive distribution) is guaranteed, as long as the multivariate linear mixed model (5.17) holds. However, it cannot handle the missingness of categorical variables.

5.8.2 Imputation by Chained Equations

For a longitudinal study with a fixed-time design, the imputation could be performed on a ''wide format'' of the data set, meaning that each subject takes up one row. The outcome and time-varying covariates, which are observed at different but fixed time points, are treated as different variables (columns). In this case, the multiple imputation procedure is the same as that in cross-sectional studies. The outcome variable at one time point could predict the outcome at another time point in the imputation model, which reflects the within-subject dependence. If a longitudinal study does not have the fixed-time design, the data set can only be recorded in a ''long format''; that is, each row corresponds to one observation of a subject and each subject contributes multiple rows. The NACC UDS data set is such an example, since the participants visited the clinics at different times. Then the imputation model should distinguish the subject-level and observation-level variables. The general framework of chained equations described in Section 2.3.2 still works. But for observational-level variables, the conditional distribution $P(Y_1^{\text{mis}}|Y_1^{\text{obs}}, Y_2^{(t)}, \ldots, Y_p^{(t)}, \theta_1^{*(t+1)})$ should specify the clustering structure, often by including random effects. For subject-level variables, the imputation should be performed only once for each subject. We will demonstrate these in detail with the analysis of NACC UDS data.

Beunckens et al. (2008) compared the MI and IPWGEE approaches in analyzing binary longitudinal data with dropout. They found that in finite sample settings, MI is less biased and more precise. In addition, MI is not very sensitive to the misspecification of the imputation model. The efficiency issue of IPWGEE could potentially be addressed by augmented IPW estimators, as we will introduce in the next section. However, the implementation of the augmented IPW estimators may be quite difficult, especially when the missingness involves both the outcome and the covariates. In these cases, the MI approach is generally recommended.

5.8.3 Example: NACC UDS Data

The NACC UDS example was introduced in Section 5.2, and we now demonstrate the use of `R` library `mice` to impute and analyze this example. Recall that our aim is to explore the risk factors that affect the decline of the MMSE score. Let Y_{ij} be the MMSE score for subject i at the jth time t_{ij}; let X_{ij} be the vector of covariates, including baseline age, gender, education, HIS, and APOE e4. Using generalized estimating equations, we can fit the following regression model:

$$E\left(Y_{ij}|X_{ij}, t_{ij}\right) = \theta_0 + \theta_1 t_{ij} + \theta_2' X_{ij} + \theta_3' X_{ij} \times t_{ij}. \tag{5.18}$$

The interaction term is of primary interest, which is interpreted as the increase in the average rate of MMSE decline per unit increase in X_{ij}.

Four variables are subject to missingness: the outcome, MMSE score, is a continuous-scale observation-level variable; HIS is an ordinal observation-level variable; APOE e4 is a binary subject-level variable; and education is an ordinal subject-level variable. Since the interaction terms are included in the analysis model (5.18), it is recommended that they should also be included in the imputation model.

We conduct the multiple imputation in several steps:

1. Prepare the data set for imputation by appending the transformation of variables, dummy variables, and interaction terms. Initialize the imputation parameters by ‘‘dry run’’ with the maximum number of iteration set to 0.
2. Specify the imputation methods. A two-level linear mixed effects model (`2l.norm` in the R package `mice`) is used to impute MMSE. MMSE is bounded between 0 and 30, whereas the `2l.norm` method draws posterior samples from an unbounded normal distribution. Therefore, we transform the MMSE score as logit ((MMSE + 0.5)/31) to force it unbounded and impute the transformed score. The back-transformed original score is generated by a ‘‘passive imputation.’’ The passive imputation means that the imputed values

are computed from other imputed variables, as opposed to be generated from an imputation model. For the ordinal HIS, `mice` cannot handle the two-level clustering, so we can choose either the predictive mean matching (`pmm`) or the ordinal logit model (`polr`) method, both of which ignore the longitudinal structure. In our example, we use the `pmm` method. For imputing education and APOE e4, we choose the predictive mean matching at the subject level only (`2lonly.pmm` in the R package `mice`). The observation-level variables are aggregated within each cluster to serve as covariates in the imputation model. All the dummy variables and interaction terms that involve missing data were passively imputed.

3. Specify the prediction matrix of the imputation model. Each row of the prediction matrix corresponds to an imputation model for a variable, where "1" indicates the column variable is used as a predictor, whereas 0 means it is not used. In the two-level imputation methods, "−2" indicates the clustering variable and "2" indicates the random effects. Passive imputation overrules the predictor matrix elements, so the rows of passively imputed variables can be set as any value.
4. Specify the visit scheme, that is, the order of imputation. We need to keep the passive imputations synchronized with the imputed variables. In our case, the imputation of `mmse2`, `education`, `apoee4`, and `hachin` should be updated first, and then the transformations, dummy variables, and interactions are generated.
5. Call the `mice` function to impute the data set.

We also try to impute MMSE using the predictive mean matching method and see whether the result changes as we ignore the within-cluster correlation. Two imputation models are implemented for $m = 20$ times. We set $m = 20$ mainly for diagnostic purposes. The imputed values are plotted in Figure 5.2, which shows that the 20 imputations are well mixed indicating healthy convergence.

We can now conduct the analysis for the 20 imputed data sets. The combination is performed using Rubin's rule. The analysis results are shown in Table 5.11. CC refers to the complete-case analysis, imputation 1 uses `2l.norm` for MMSE, and imputation 2 uses `pmm` for MMSE. We can see that both imputation methods give similar results, and the results of CC might be a bit biased with larger standard errors. Since baseline age is centered at 75 years, the coefficient for "year since aMCI" is interpreted as the average rate of decline per year (about 0.3 point decrease every year) for a female aged 75 at the baseline, with 0–12 years of education, 0 point HIS, and no APOE e4 allele. Imputation model 2 identifies all five risk factors to affect the level of MMSE; three of the risk factors significantly affect the rate of MMSE decline, namely, age, APOE e4, and education. For example, a subject with APOE e4 allele has about 0.44 point more decrease in MMSE every year compared with a subject without APOE e4 allele.

Although in this example ignoring the clustering in the imputation model does not seem to affect the final results, this is not warranted in other settings. Future

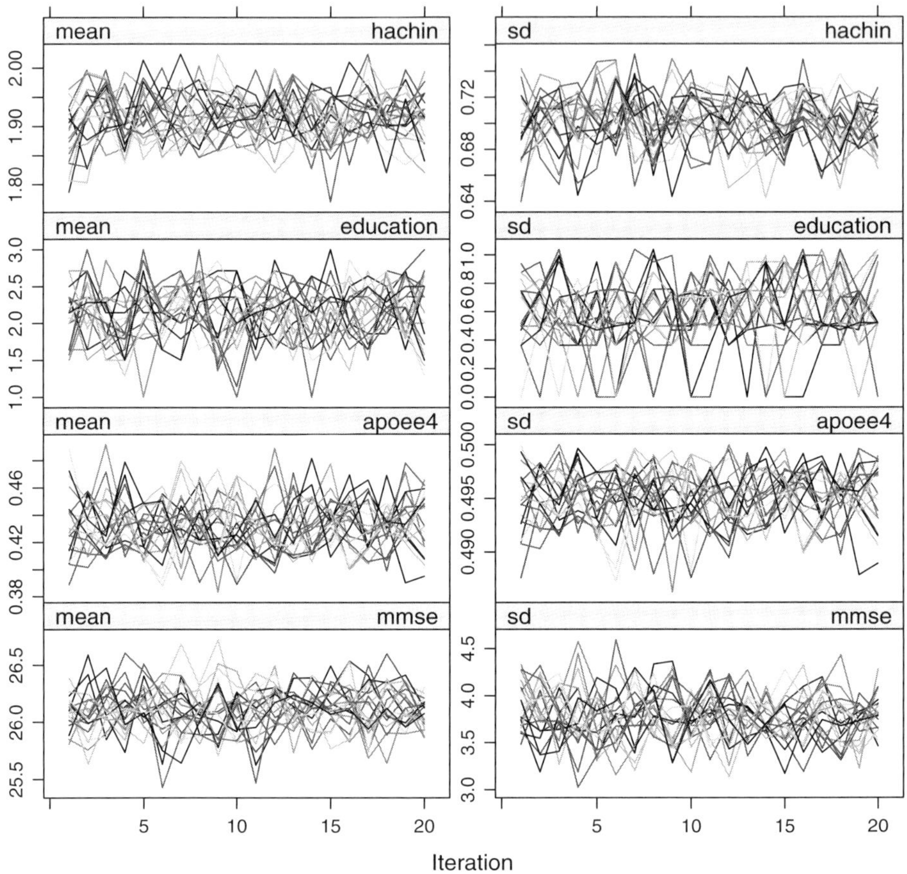

FIGURE 5.2 Diagnostic plot for the imputed `hachin`, `education`, `mmse2`, and `apoee4`.

research is needed to explore the effect of misspecifying the imputation model. The linear mixed effects model is the only available build-in function for the two-level imputation in `mice`, but the multivariate normality assumption is quite strong in practice. If more complicated imputation models are to be used in `mice`, the readers should refer to van Buuren and Groothuis-Oudshoorn (2011) for the user-defined imputation functions.

5.8.4 Example: IMPACT Study

5.8.4.1 Multiple Imputation with Monotone Missingness With monotone missingness, the joint imputation method greatly simplifies in terms of the computation load. The basic idea is that missing data are imputed sequentially, from the

TABLE 5.11 Analysis Results of NACC UDS Data

		Imputation 1	Imputation 2	CC
(Intercept)		26.51 (0.15)	26.51 (0.15)	26.55 (0.18)
Age	Per 10 years	**−0.39 (0.08)**	**−0.37 (0.07)**	**−0.33 (0.09)**
Gender	Male	**−0.40 (0.10)**	**−0.36 (0.10)**	**−0.44 (0.13)**
APOE e4		**−0.48 (0.13)**	**−0.49 (0.12)**	**−0.45 (0.13)**
Education	13–16 years	**1.31 (0.14)**	**1.35 (0.14)**	**1.25 (0.17)**
	17+ years	**1.71 (0.15)**	**1.74 (0.14)**	**1.81 (0.18)**
HIS	1 point	−0.12 (0.11)	−0.11 (0.11)	−0.09 (0.13)
	2+ points	**−0.33 (0.16)**	**−0.32 (0.16)**	−0.10 (0.19)
Year since aMCI		**−0.26 (0.10)**	**−0.29 (0.11)**	**−0.24 (0.12)**
Age:year	Per 10 years	−0.06 (0.05)	**−0.14 (0.06)**	**−0.14 (0.06)**
Gender:year	Male	−0.02 (0.07)	−0.04 (0.07)	−0.05 (0.08)
APOE e4:year		**−0.42 (0.08)**	**−0.44 (0.09)**	**−0.53 (0.09)**
Education:year	13–16 years	**−0.21 (0.09)**	**−0.22 (0.09)**	−0.18 (0.10)
	17+ years	−0.14 (0.09)	−0.13 (0.10)	−0.12 (0.11)
HIS:year	1 point	0.09 (0.08)	0.09 (0.09)	0.10 (0.10)
	2+ points	0.02 (0.11)	−0.03 (0.12)	−0.10 (0.14)

variable with the least proportion of missingness to the one with the most missingness. While imputing for longitudinal data, the within-subject correlation needs to be taken into account as well. In general, we recommend multivariate imputation methods such as multivariate normal or chained equations for this type of imputation task. In the Stata code, given in Appendix 5.A, baseline covariates such as age, gender, study site, and so on are first used to impute the SCL-20 score at baseline using the linear regression model, and then baseline covariates plus the SCL-20 score at baseline are used to impute the SCL-20 score at month 3, and so on until missing SCL-20 scores at month 24 are imputed. In the following code, 10 imputed data sets are generated. Estimation is performed on each imputed data set, and the results are then pooled together to produce the final result. Because the analysis model examines the interaction between time and treatment, we conduct the imputation for the subjects in the intervention and control arms separately.

The results are listed in Table 5.12. The estimated intervention effect (*group* × *period*) is slightly smaller than that estimated based on available data analysis.

5.8.4.2 Multiple Imputation Based on Multivariate Normal Distribution

For a continuous longitudinal outcome with missing values, another approach for missing data imputation is to first assume a multivariate normal distribution on the outcome vector and then to proceed with the imputation using the likelihood approach. The details have been provided in Schafer (1997b). Also note that this imputation approach works for arbitrary missing data patterns. For categorical outcome with missing values, latent variable models could be employed with this imputation approach.

TABLE 5.12 Estimates Based on the Multiple Imputation with Monotone Missingness

Parameter	Estimate	Robust SE	95% CI
β_1	1.686	0.022	(1.644, 1.729)
β_2	−0.335	0.022	(−0.379, −0.291)
β_3	−0.003	0.030	(−0.056, 0.062)
β_4	−0.288	0.031	(−0.350, −0.227)
$\sqrt{g_{11}} = \text{SD}(b_{1i})$	0.453	0.014	(0.426, 0.482)
$\sqrt{g_{22}} = \text{SD}(b_{2i})$	0.415	0.018	(0.382, 0.451)
$\text{Corr}(b_{1i}, b_{2i})$	−0.202	0.048	(−0.293, −0.108)

Table 5.13 shows the results of the multiple imputation method based on MVN. It can be seen that the results are close to those based on the multiple imputation with monotone missingness (Table 5.12).

5.8.4.3 Multiple Imputation Based on Chained Equations We next apply the multiple imputation with chained equations (MICE) approach to the IMPACT data set with monotone missingness. The results are shown in Table 5.14, from which it can be seen that the results are similar to the results based on the above two multiple imputation methods. It is worth noting that MICE applies to general missing data patterns. The results are very similar to those from both monotone MI and MVN MI.

5.8.4.4 Multiple Imputation with Chained Equations for a General Missing Data Pattern We now illustrate MICE when there are both the missing outcome and missing covariates with a general missing data pattern. In The IMPACT data, not only there are missing values in the SCL-20 score, there are also missing values in baseline number of chronic diseases (numdis1), preference on depression treatment (pref00), race (white), baseline general health status (ghlth00), and income (inc400). It is conceivable that we could do a better imputation of the

TABLE 5.13 Estimates Based on the Multiple Imputation with Multivariate Normal Distribution

Parameter	Estimate	Robust SE	95% CI
β_1	1.686	0.022	(1.644, 1.729)
β_2	−0.338	0.023	(−0.383, −0.293)
β_3	−0.003	0.030	(−0.056, 0.062)
β_4	−0.286	0.032	(−0.348, −0.224)
$\sqrt{g_{11}} = \text{SD}(b_{1i})$	0.454	0.014	(0.427, 0.482)
$\sqrt{g_{22}} = \text{SD}(b_{2i})$	0.422	0.019	(0.386, 0.462)
$\text{Corr}(b_{1i}, b_{2i})$	−0.207	0.049	(−0.301, −0.109)

TABLE 5.14 Estimates Based on Multiple Imputation with Chained Equations

Parameter	Estimate	Robust SE	95% CI
β_1	1.686	0.022	(1.644, 1.729)
β_2	−0.335	0.022	(−0.379, −0.291)
β_3	−0.003	0.030	(−0.056, 0.062)
β_4	−0.288	0.031	(−0.350, −0.227)
$\sqrt{g_{11}} = \mathrm{SD}(b_{1i})$	0.453	0.014	(0.426, 0.482)
$\sqrt{g_{22}} = \mathrm{SD}(b_{2i})$	0.415	0.018	(0.382, 0.451)
$\mathrm{Corr}(b_{1i}, b_{2i})$	−0.202	0.048	(−0.293, −0.108)

TABLE 5.15 Estimates Based on Multiple Imputation with Chained Equations for a General Missing Data Pattern

Parameter	Estimate	Robust SE	95% CI
β_1	1.672	0.020	(1.632, 1.712)
β_2	−0.315	0.022	(−0.358, −0.273)
β_3	−0.010	0.029	(−0.046, 0.066)
β_4	−0.289	0.031	(−0.349, −0.227)
$\sqrt{g_{11}} = \mathrm{SD}(b_{1i})$	0.448	0.014	(0.422, 0.476)
$\sqrt{g_{22}} = \mathrm{SD}(b_{2i})$	0.419	0.019	(0.384, 0.457)
$\mathrm{Corr}(b_{1i}, b_{2i})$	−0.206	0.047	(−0.295, −0.112)

missing outcome if we can gather more information on these important covariates. Specifically, we used logistic regression to impute race, ordinal logistic regression for treatment preference and general health status given they are multilevel categorical variables, Poisson regression model for a number of chronic conditions, predictive mean matching for income, and finally linear regression for SCL-20 scores. The results are shown in Table 5.15. It can be seen that for the IMPACT study, the results using all four multiple imputation methods generally agree.

From Tables 5.5–5.7, 5.9, 5.10, and 5.12–5.15, it is seen that for the IMPACT study, all the missing data methods show similar results, except the IPWGEE1. In general, we recommend the mixed effects model and multiple imputation methods for longitudinal data analysis with missing data. This is also the recommendation given by Schafer and Graham (2002).

5.9 BAYESIAN INFERENCE

The Bayesian approach for a cross-sectional regression model has been discussed in Chapter 4. Under the assumption of MAR, the missing mechanism model can be ignored. So, the data augmentation procedures can be used to estimate the posterior distribution of the regression parameters. The same model framework applies to longitudinal studies, except that in the missing data model, the within-cluster correlation needs to be characterized by mixed effects models.

The outcome vector $\boldsymbol{Y}_i$ is partitioned into $(\boldsymbol{Y}_i^{\text{obs}}, \boldsymbol{Y}_i^{\text{mis}})$. The covariate matrix $\boldsymbol{X}_i$ can also be partitioned as $(\boldsymbol{X}_i^{(1)}, \boldsymbol{X}_i^{(2)})$, where $\boldsymbol{X}_i^{(1)}$ includes the covariates subject to missingness and $\boldsymbol{X}_i^{(2)}$ includes the completely observed covariates. For simplicity, assume that only one covariate is subject to missingness, so $\boldsymbol{X}_i^{(1)}$ has one column. If $\boldsymbol{X}_i^{(1)}$ is a time-varying covariate, we can further partition $\boldsymbol{X}_i^{(1)}$ to be $(\boldsymbol{X}_i^{(1),\text{obs}}, \boldsymbol{X}_i^{(1),\text{mis}})$. And we use $\boldsymbol{X}_i^{\text{obs}}$ to denote $\{\boldsymbol{X}_i^{(1),\text{obs}}, \boldsymbol{X}_i^{(2)}\}$, and $\boldsymbol{X}_i^{\text{mis}}$ to denote $\boldsymbol{X}_i^{(1),\text{mis}}$. Let $\boldsymbol{\theta}$ be the vector of unknown parameters in the regression model of $\boldsymbol{Y}_i$ versus $\boldsymbol{X}_i$, as well as the nuisance parameters used to specify the missing covariate model, $\boldsymbol{X}_i^{(1)}|\boldsymbol{X}_i^{(2)}$.

Estimation of the posterior distribution $f(\boldsymbol{\theta}|\boldsymbol{Y}_i^{\text{obs}}, \boldsymbol{X}_i^{(1),\text{obs}}, \boldsymbol{X}_i^{(2)})$ is of our primary interest. The data augmentation procedures can be used to draw the target posterior distribution in two iteration steps: imputation step and posterior step. In the $(t+1)$th iteration, the imputation step draws the missing data from the conditional distribution:

$$f\left(\boldsymbol{Y}_i^{\text{mis}}, \boldsymbol{X}_i^{\text{mis}}|\boldsymbol{Y}_i^{\text{obs}}, \boldsymbol{X}_i^{\text{obs}}, \boldsymbol{\theta}^{(t)}\right). \tag{5.19}$$

The imputation step results in a complete data set with no missing values. Then, the posterior step takes a random draw from the posterior conditional distribution:

$$\left\{\prod_{i=1}^{N} L\left(\boldsymbol{\theta}|\boldsymbol{Y}_i^{\text{obs}}, \boldsymbol{X}_i^{\text{obs}}, \boldsymbol{Y}_i^{\text{mis},(t+1)}, \boldsymbol{X}_i^{\text{mis},(t+1)}\right)\right\} \times p(\boldsymbol{\theta}),$$

where $L(\boldsymbol{\theta}|\boldsymbol{Y}_i^{\text{obs}}, \boldsymbol{X}_i^{\text{obs}}, \boldsymbol{Y}_i^{\text{mis},(t+1)}, \boldsymbol{X}_i^{\text{mis},(t+1)})$ is the complete-data likelihood for subject i and $p(\boldsymbol{\theta})$ is the prior distribution for the parameters.

The posterior step is simply a standard Bayesian analysis without missing data, and the posterior sample can be drawn using Gibbs sampling. The difficulty mainly lies in the imputation step. Note that

$$\begin{aligned}
& f\left(\boldsymbol{Y}_i^{\text{mis}}, \boldsymbol{X}_i^{\text{mis}}|\boldsymbol{Y}_i^{\text{obs}}, \boldsymbol{X}_i^{\text{obs}}, \boldsymbol{\theta}^{(t)}\right) \\
\propto\ & f\left(\boldsymbol{Y}_i^{\text{mis}}, \boldsymbol{X}_i^{\text{mis}}, \boldsymbol{Y}_i^{\text{obs}}, \boldsymbol{X}_i^{\text{obs}}|\boldsymbol{\theta}^{(t)}\right) \\
=\ & f\left(\boldsymbol{Y}_i, \boldsymbol{X}_i^{(1)}, \boldsymbol{X}_i^{(2)}|\boldsymbol{\theta}^{(t)}\right) \\
=\ & f\left(\boldsymbol{Y}_i|\boldsymbol{X}_i, \boldsymbol{\theta}^{(t)}\right) f(\boldsymbol{X}_i^{(1)}|\boldsymbol{X}_i^{(2)}, \boldsymbol{\theta}^{(t)}) f\left(\boldsymbol{X}_i^{(2)}|\boldsymbol{\theta}^{(t)}\right) \\
\propto\ & f\left(\boldsymbol{Y}_i|\boldsymbol{X}_i, \boldsymbol{\theta}^{(t)}\right) f\left(\boldsymbol{X}_i^{(1)}|\boldsymbol{X}_i^{(2)}, \boldsymbol{\theta}^{(t)}\right).
\end{aligned}$$

Here, the density of $\boldsymbol{X}_i^{(2)}$ can be ignored as it does not include any missing values. It suffices to specify the full data models $\boldsymbol{Y}_i|\boldsymbol{X}_i$ and $\boldsymbol{X}_i^{(1)}|\boldsymbol{X}_i^{(2)}$. The conditional distribution of $\boldsymbol{Y}_i|\boldsymbol{X}_i$ is given by the regression model of interest, which is often a generalized linear mixed effects model; the conditional distribution $\boldsymbol{X}_i^{(1)}|\boldsymbol{X}_i^{(2)}$ can be specified by another regression model.

Although the Bayesian inference can be outlined simply in two iterative procedures, the computation load could be quite intensive for several reasons. First, if the outcome $\boldsymbol{Y}_i$ is categorical, the marginal likelihood involves intractable integrals. So additional steps are taken to "augment" the random effects as if they were the missing data. Second, in practice, multiple covariates could be subject to missingness, so the imputation step could involve calculation of the quite complicated multivariate distribution of the missing data. It could be time-consuming to draw imputation samples.

5.10 OTHER APPROACHES

Many other missing data methods have been proposed in the literature. However, they are not implemented in standard statistical packages and are less commonly used in practice. We briefly review some estimating equation approaches in this section and provide references for the interested readers. Particularly, we first discuss a mean score imputation approach, and then review several doubly robust methods.

5.10.1 Imputing Estimating Equations

Paik (1997) proposed an imputation approach that could handle the dropout of the outcome variable. The approach is based on generalized estimating equations, and the equations with missing observations are replaced by their conditional expectations. This idea is the same as the "mean score imputation" in the cross-sectional setting (Chapter 4). Two important assumptions are as follows: (a) The longitudinal observations are all scheduled at fixed time points, $j = 1, \ldots, M$. (b) The dropout process depends only on observed variables, that is, the MAR dropout process.

Let $\boldsymbol{Y}_i^{\mathrm{F}} = (Y_{i1}, \ldots, Y_{in_i}, \ldots, Y_{iM})'$ be the full vector of outcome variables for subject i and $\boldsymbol{X}_i^{\mathrm{F}} = (\boldsymbol{X}_{i1}, \ldots, \boldsymbol{X}_{in_i}, \ldots, \boldsymbol{X}_{iM})'$ be the full matrix of corresponding covariates. The primary interest is the regression model of $\boldsymbol{\mu}_i^{\mathrm{F}} \equiv E(\boldsymbol{Y}_i^{\mathrm{F}}|\boldsymbol{X}_i^{\mathrm{F}})$. The observed data for subject i are from the first n_i visits: $\boldsymbol{Y}_i = (Y_{i1}, \ldots, Y_{in_i})'$ and $\boldsymbol{X}_i = (\boldsymbol{X}_{i1}, \ldots, \boldsymbol{X}_{in_i})'$. Let $\boldsymbol{H}_{ij} = (\boldsymbol{X}_i, Y_{i1}, \ldots, Y_{ij})$ be the history of ith subject up to time j. Let R_{ij} be the missing indicator of Y_{ij}, similar as in the previous sections.

The full data estimating equation can be written as

$$U(\boldsymbol{\beta}) = \sum_{i=1}^{N} \left(\boldsymbol{D}_i^{\mathrm{F}}\right)' \left(\boldsymbol{V}_i^{\mathrm{F}}\right)^{-1} (\boldsymbol{Y}_i^{\mathrm{F}} - \boldsymbol{\mu}_i^{\mathrm{F}}) = 0,$$

where $D_i^{\mathrm{F}} = \partial \boldsymbol{\mu}_i^{\mathrm{F}} / \partial \boldsymbol{\beta}$ and V_i^{F} is a working covariance matrix for $\boldsymbol{Y}_i^{\mathrm{F}}$. Paik's approach uses $\tilde{Y}_{ij} \equiv E\left(Y_{ij} | \boldsymbol{H}_{in_i}, R_{ij} = 0\right)$ to replace those missing Y_{ij} ($j > n_i$). Then, the imputed estimating equation becomes

$$\tilde{U}(\boldsymbol{\beta}) = \sum_{i=1}^{N} \left(D_i^{\mathrm{F}}\right)' \left(V_i^{\mathrm{F}}\right)^{-1} (\tilde{\boldsymbol{Y}}_i^{\mathrm{F}} - \boldsymbol{\mu}_i^{\mathrm{F}}) = 0, \tag{5.20}$$

where $\tilde{\boldsymbol{Y}}_i^{\mathrm{F}} = (Y_{i1}, \ldots, Y_{in_i}, \tilde{Y}_{i,n_i+1}, \ldots, \tilde{Y}_{iM})'$. The expectation of equation (5.20) is 0, which leads to a consistent estimator of $\boldsymbol{\beta}$. Now the crucial step is to estimate the conditional expectation $\tilde{Y}_{ij}$.

Note that under MAR dropout,

$$E(Y_{ij} | \boldsymbol{H}_{i,j-1}, R_{i,j-1} = 1, R_{ij} = 0) = E(Y_{ij} | \boldsymbol{H}_{i,j-1}, R_{i,j-1} = 1, R_{ij} = 1). \tag{5.21}$$

So, a regression model can be fitted to the observed data, in order to estimate the first missing observation after dropout. For the second missing observation after dropout, we have

$$E(Y_{ij} | \boldsymbol{H}_{i,j-2}, R_{i,j-2} = 1, R_{i,j-1} = 0) = E(Y_{ij} | \boldsymbol{H}_{i,j-2}, R_{i,j-1} = 1). \tag{5.22}$$

Hence, another regression model can be fitted to estimate the conditional expectation $E(Y_{ij} | \boldsymbol{H}_{i,j-2}, R_{i,j-1} = 1)$. Now the regression model includes not only the observed data (those with $R_{i,j-1} = 1$ and $R_{i,j} = 1$) but also the previously imputed observations (those with $R_{i,j-1} = 1$ and $R_{i,j} = 0$). In other words, in fitting the regression models to impute the second missing observation, the imputed value of the first missing observation is also used as the outcome. If the previous imputation is consistent, the current regression model is also consistent. The same techniques can be repeated for the third potential observation, and so on, until all the missing data are replaced by the fitted conditional expectations. Eventually, the consistency of estimating $\tilde{Y}_{ij}$ is guaranteed in this procedure, if all the imputation models are correctly specified.

The whole procedure can be illustrated in Figure 5.3. The rows represent the missing patterns and the columns represent the observation times. The shaded cells are the missing observations that are to be imputed. We put each one of the subjects into a row according to their dropout times. The n_i observations for subject i enter the n_i cells in its row. The first diagonal of missing values is imputed in the first place, then the second diagonal, and so on. For the imputation of a cell c, the regression model uses all the cells in the same column but below cell c. For example, when imputing cell (3, 4) in the first diagonal, a regression model is fitted to observations in cells (4, 4) and (5, 4)—both are observed cells. When imputing cell (2, 4) in the second diagonal,

Missing pattern	$j=1$	$j=2$	$j=3$	$j=4$	$j=5$	
1						
2				(2, 4)		Fourth diagonal
3				(3, 4)		Third diagonal
4				(4, 4)		Second diagonal
5				(5, 4)		First diagonal

FIGURE 5.3 Illustration of the imputation procedures.

not only cells $(4, 4)$ and $(5, 4)$ but also the imputed cell $(3, 4)$ will enter the regression model as outcomes. The design matrix, however, is the history up to the second time point ($\boldsymbol{H}_{i,j-2}$ in (5.22) is now $\boldsymbol{H}_{i2}$), which does not involve any imputed cell or missing cell. The sequential imputation of the diagonals will eventually generate the complete data, and the estimation of the regression coefficients is followed by solving equation (5.20).

The explicit variance formula is given in Theorem 1 and Appendix B of Paik (1997). The variance indeed comes from two sources: one is from the estimating equation (5.20) while pretending all the imputed data were observed; the other one is from estimating the additional regression models for imputation.

In Paik's mean score imputation, the missing variables are replaced with the fitted values because the full data estimating equations are linear in the missing Y_{ij}. In the multiple imputation, as a comparison, the missing Y_{ij} are replaced with a random draw from the posterior predictive distribution. Under the correct imputation model, the multiply imputed values $\hat{Y}_{ij}^{(1)}, \ldots, \hat{Y}_{ij}^{(M)}$ would be close on average to the fitted value $\tilde{Y}_{ij}$ in the regression model, that is, Paik's mean score:

$$\frac{1}{K}\sum_{k=1}^{K} \hat{Y}_{ij}^{(k)} \approx \tilde{Y}_{ij},$$

when M gets larger. In a sense, Paik's mean score imputation is the limit situation of the multiple imputation. As shown in the simulation results of Paik (1997), the results of Paik's imputation and multiple imputation are similar in many settings.

5.10.2 Doubly Robust Estimation

Seaman and Copas (2009) proposed a doubly robust estimating equation approach to handle the monotone dropout of the outcome variable. Their approach is closely related to Paik's mean score imputation. The idea is to augment the complete-case IPW (IPWCC) estimating equations. The augmentation term is the expected complete-data estimating equation, which can be computed in the same way as in Paik (1997).

We continue to use all the notations in the previous section. Note that the IPWCC estimating equations are given by

$$\sum_{i=1}^{N} \frac{R_{iM}}{\pi_{iM}} U_i(\boldsymbol{\beta}) = \sum_{i=1}^{N} \frac{R_{iM}}{\pi_{iM}} \left(\boldsymbol{D}_i^{\mathrm{F}}\right)' \left(\boldsymbol{V}_i^{\mathrm{F}}\right)^{-1} (\boldsymbol{Y}_i^{\mathrm{F}} - \boldsymbol{\mu}_i^{\mathrm{F}}).$$

The IPWCC equations ignore the available data on individuals who drop out before the end of study (time M), and hence are often quite inefficient. One way to improve the efficiency is to add an additional term A, which is a function of both the available data and the parameter vector $\boldsymbol{\beta}$ and satisfies $E(A|\boldsymbol{H}_{iM}) = 0$. The augmented IPWCC (AIPWCC) equations are given by

$$\sum_{i=1}^{N} \left\{ \frac{R_{iM}}{\pi_{iM}} U_i(\boldsymbol{\beta}) + A_i \right\} = 0.$$

Let C_{ij} be the indicator function, which is equal to 1 if $R_{ij} = 1$ and $R_{i,j+1} = 0$, and 0 otherwise. One choice of the augmentation function A_i is

$$A_i = \sum_{j=1}^{M-1} \left(\frac{C_{ij} - \left(1 - \lambda_{i,j+1}\right) R_{ij}}{\pi_{i,j+1}} \right) E\left\{U_i(\boldsymbol{\beta}) \,|\, \boldsymbol{H}_{ij}, R_{ij} = 1\right\}, \tag{5.23}$$

where $\lambda_{i,j+1} = \Pr(R_{i,j+1} = 1|R_{ij} = 1, \boldsymbol{H}_{ij})$ is defined in Section 5.6.1. Note that the expectation is taken with respect to future outcome variables, given the history up to time j and the patient not having dropped out at time j.

As $U_i(\boldsymbol{\beta})$ is linear in all the outcome variables Y_{ij}, it suffices to compute the expectation $E\left(Y_{im}|\boldsymbol{H}_{ij}, R_{ij} = 1\right)$. If $m \leq j$, the expectation is just the observed value Y_{im}. Otherwise, the Paik's imputation procedures are used to compute the fitted mean.

The AIPWCC estimator is doubly robust, meaning that it is consistent when either the dropout model or the imputation model is misspecified (but not both). If both models are correct, the choice of A_i in (5.23) is optimal in terms of the best efficiency for $\boldsymbol{\beta}$.

The variance estimator for $\boldsymbol{\beta}$ has the sandwich form, and the formula is given in Section 4 of Seaman and Copas (2009). When the dropout model and the imputation model are both correct, it is valid to ignore the uncertainty of the nuisance parameters. In other words, using the true dropout and imputation models is approximately the same as using the estimated models.

5.10.3 Missing Outcome and Covariates

The literature on missing data methodologies mainly focuses on either missing response or missing covariates. In practice, it is often inevitable that data are incomplete for both the longitudinal outcome and some important time-varying covariates. We will now discuss some methods that address this issue (Chen et al., 2010; Chen and Zhou, 2011). One of these methods was based on a weighted GEE approach that extended the results of Robins et al. (1995), whereas the other one further developed a doubly robust estimator when the outcome was binary.

We need to redefine some of the notations. Consider a data set with N subjects indexed by i, and subject i is to be examined at time points $j = 1, \dots, n_i$. Let $\boldsymbol{Y}_i = (Y_{i1}, \dots, Y_{in_i})'$ be the outcome vector, $\boldsymbol{X}_i = (X_{i1}, \dots, X_{in_i})'$ be the covariate vector that is subject to missingness, and $\boldsymbol{Z}_i = (\boldsymbol{Z}_{i1}, \dots, \boldsymbol{Z}_{in_i})'$ be the matrix of covariates that are fully observed. For simplicity, we assume that only one covariate is missing. The model of interest is the conditional mean of $\boldsymbol{Y}_i$, that is, $\boldsymbol{\mu}_i \equiv (\mu_{i1}, \dots, \mu_{in_i})' = E(\boldsymbol{Y}_i | \boldsymbol{X}_i, \boldsymbol{Z}_i)$. Suppose that the marginal regression model is given by

$$g(\mu_{ij}) = X_{ij}\beta_x + \boldsymbol{Z}'_{ij}\boldsymbol{\beta}_z,$$

for $i = 1, \dots, N$, $j = 1, \dots, n_i$, where g is a known link function. Let R_{ij} be the indicator of the missingness patterns for the pair (Y_{ij}, X_{ij}):

$$R_{ij} = \begin{cases} 0, & \text{if both } Y_{ij} \text{ and } X_{ij} \text{ are missing,} \\ 1, & \text{if } Y_{ij} \text{ is missing and } X_{ij} \text{ is observed,} \\ 2, & \text{if } Y_{ij} \text{ is observed and } X_{ij} \text{ is missing,} \\ 3, & \text{if both } Y_{ij} \text{ and } X_{ij} \text{ are observed.} \end{cases}$$

Let $\boldsymbol{R}_i = \left(R_{i1}, \dots, R_{in_i}\right)'$ be the full missingness record for subject i and $\bar{\boldsymbol{R}}_{ij} = \{R_{i1}, \dots, R_{i,j-1}\}$ be the missingness record before time j. Suppose that the baseline outcome and covariates are always observed, that is, $R_{i1} \equiv 3$.

Chen and Zhou (2011) proposed to model each R_{ij} conditional on the past missingness profile, $P(R_{ij} | \bar{\boldsymbol{R}}_{ij}, \boldsymbol{Y}_i, \boldsymbol{X}_i, \boldsymbol{Z}_i)$, to obtain the joint probability of $\boldsymbol{R}_i$. The joint probability of $\boldsymbol{R}_i$ is then expressed as

$$P(\boldsymbol{R}_i | \boldsymbol{Y}_i, \boldsymbol{X}_i, \boldsymbol{Z}_i) = \prod_{j=2}^{n_i} P(R_{ij} | \bar{\boldsymbol{R}}_{ij}, \boldsymbol{Y}_i, \boldsymbol{X}_i, \boldsymbol{Z}_i).$$

Under the MAR assumption, it is natural to assume that

$$P(R_{ij} | \bar{\boldsymbol{R}}_{ij}, \boldsymbol{Y}_i, \boldsymbol{X}_i, \boldsymbol{Z}_i) = P(R_{ij} | \bar{\boldsymbol{R}}_{ij}, \bar{\boldsymbol{Y}}^{\text{o}}_{ij}, \bar{\boldsymbol{X}}^{\text{o}}_{ij}, \boldsymbol{Z}_i),$$

where $\bar{\boldsymbol{Y}}^{\text{o}}_{ij}$ and $\bar{\boldsymbol{X}}^{\text{o}}_{ij}$, respectively, denote the history of observed component of $\boldsymbol{Y}_i$ and $\boldsymbol{X}_i$ before time j. Let $\lambda_{ijk} \equiv P(R_{ij} = k | \bar{\boldsymbol{R}}_{ij}, \bar{\boldsymbol{Y}}^{\text{o}}_{ij}, \bar{\boldsymbol{X}}^{\text{o}}_{ij}, \boldsymbol{Z}_i)$. A multinomial logit model can then be assumed for R_{ij}:

$$\log \frac{\lambda_{ijk}}{\lambda_{ij0}} = W'_{ijk}\alpha_k, \quad \text{for } k = 1, 2, 3,$$

where W_{ijk} is a subset of $\left\{ \bar{\boldsymbol{R}}_{ij}, \bar{\boldsymbol{Y}}^{\mathrm{o}}_{ij}, \bar{\boldsymbol{X}}^{\mathrm{o}}_{ij}, \boldsymbol{Z}_i \right\}$. Let $\pi_{ij} = P(R_{ij} = 3|\boldsymbol{Y}_i, \boldsymbol{X}_i, \boldsymbol{Z}_i)$ be the marginal probability of observing both Y_{ij} and X_{ij}, which can be written as a function of λ_{ijk} by summing over all possible combinations of $R_{i1}, \ldots, R_{i,j-1}$. The marginal probability π_{ij} is crucial to construct a weight function in the next step.

Similar to the spirit of Robins et al. (1995), a weight matrix is defined as

$$\Delta_i^* = \operatorname{diag}\left\{ \frac{I(R_{i1} = 3)}{\pi_{i1}}, \cdots, \frac{I(R_{in_i} = 3)}{\pi_{in_i}} \right\}.$$

Then, the estimating equation for $\boldsymbol{\beta}$ is obtained as

$$\sum_{i=1}^{N} D_i' V_i^{-1} \Delta_i^* (\boldsymbol{Y}_i - \boldsymbol{\mu}_i) = 0, \tag{5.24}$$

where $D_i = \partial \boldsymbol{\mu}_i / \partial \boldsymbol{\beta}$ and V_i is the working covariance matrix of Y_i. We could also express V_i as $F_i^{1/2} C_i F_i^{1/2}$, where C_i is the working correlation matrix and $F_i = \operatorname{diag}\left\{ v_{ij}, j = 1, \ldots, n_i \right\}$ describes the variance–mean relationship. Equation (5.24) is computable only when C_i is the identity matrix, that is, under working-independent correlation structure. Otherwise, as $\boldsymbol{D}_i$ may involve some unobserved covariates X_{ij}, equation (5.24) cannot be evaluated. A modification is to define a new weight matrix Δ_i:

$$\Delta_i = \left[\delta_{ijj'}\right]_{n_i \times n_i},$$

$$\delta_{ijj'} = \begin{cases} \frac{I(R_{ij}=1, R_{ij'}=3) + I(R_{ij}=3, R_{ij'}=3)}{\pi_{ijj'}}, & \text{for } j \neq j', \\ \frac{I(R_{ij}=3)}{\pi_{ij}}, & \text{for } j = j', \end{cases}$$

$$\begin{aligned} \pi_{ijj'} = {} & P(R_{ij} = 1, R_{ij'} = 3|\boldsymbol{Y}_i, \boldsymbol{X}_i, \boldsymbol{Z}_i) \\ & + P(R_{ij} = 3, R_{ij'} = 3|\boldsymbol{Y}_i, \boldsymbol{X}_i, \boldsymbol{Z}_i). \end{aligned}$$

Let $M_i = F_i^{-1/2}\left(C_i^{-1} \cdot \Delta_i\right) F_i^{-1/2}$, where $A \cdot B = \left[a_{ij} b_{ij}\right]$ denotes the Hadamard product of $n_i \times n_i$ matrices $A = \left[a_{ij}\right]$ and $B = \left[b_{ij}\right]$. Now the following modified estimating equation can be solved for $\boldsymbol{\beta}$:

$$\sum_{i=1}^{N} D_i' M_i (\boldsymbol{Y}_i - \boldsymbol{\mu}_i) = 0. \tag{5.25}$$

Equation (5.25) is the key result in Chen et al. (2010), which no longer contains the missing X_{ij}. It is also worth noting that Chen et al. (2010) adopted a slightly different approach to model the missingness mechanism. Instead of constructing a conditional model, they used the Bahadur model framework and specified the marginal missingness mechanisms for X_{ij} and Y_{ij}, as well as the pairwise correlations. Interested readers should refer to their work and Bahadur (1961) for more details.

It is well known in the literature that the inverse probability weighted estimator is usually not efficient, especially if the selection probability is low. In general, one can augment the weighted estimating equations to gain efficiency (Chen and Zhou, 2011), that is,

$$\sum_{i=1}^{N} \left\{ D_i' M_i (\boldsymbol{Y}_i - \boldsymbol{\mu}_i) + \phi_i \right\} = 0. \tag{5.26}$$

An optimal choice of ϕ_i is

$$\phi_{i,\text{opt}} = E_{\boldsymbol{Y}_i^m, \boldsymbol{X}_i^m | \boldsymbol{Y}_i^{\text{o}}, \boldsymbol{X}_i^{\text{o}}, \boldsymbol{Z}_i, \boldsymbol{R}_i} \left\{ D_i' N_i (\boldsymbol{Y}_i - \boldsymbol{\mu}_i) \right\}, \tag{5.27}$$

with $N_i \equiv F_i^{-1/2} \left\{ C_i^{-1} \cdot \left(\mathbf{1}_i - \Delta_i \right) \right\} F_i^{-1/2}$, where $\mathbf{1}_i$ is an $n_i \times n_i$ square matrix of 1's.

In order to evaluate the conditional expectation (5.27), one needs to specify two missing data models of $\boldsymbol{Y}_i^m | \boldsymbol{Y}_i^{\text{o}}, \boldsymbol{X}_i, \boldsymbol{Z}_i$ and $\boldsymbol{X}_i^m | \boldsymbol{Y}_i^{\text{o}}, \boldsymbol{X}_i^{\text{o}}, \boldsymbol{Z}_i$. The missing data model for $\boldsymbol{Y}_i$ is specified from the Bahadur model, when the outcome is binary. Let $Y_{ij}^* = \left(Y_{ij} - \mu_{ij} \right) / \sqrt{\mu_{ij}(1 - \mu_{ij})}$ and

$$\begin{aligned}
\rho_{i,jk} &= \text{corr}(Y_{ij}^*, Y_{ik}^*) = E(Y_{ij}^* Y_{ik}^*), \\
\rho_{i,jkl} &= E(Y_{ij}^* Y_{ik}^* Y_{il}^*), \\
\vdots \quad & \vdots \quad \vdots \\
\rho_{i,1\cdots n_i} &= E\left(Y_{i1}^* \cdots Y_{in_i}^* \right).
\end{aligned}$$

Then, the Bahadur model specifies the joint probability density function for $\boldsymbol{Y}_i$ as

$$\begin{aligned}
P(\boldsymbol{Y}_i | \boldsymbol{X}_i, \boldsymbol{Z}_i) &= \left\{ \prod_{j=1}^{n_i} \mu_{ij}^{Y_{ij}} (1 - \mu_{ij})^{1 - Y_{ij}} \right\} \\
&\times \left(1 + \sum_{j<k} \rho_{i,jk} Y_{ij}^* Y_{ik}^* + \sum_{j<k<l} \rho_{i,jkl} Y_{ij}^* Y_{ik}^* Y_{il}^* + \cdots + \rho_{i,1\cdots n_i} Y_{i1}^* \cdots Y_{in_i}^* \right).
\end{aligned}$$

Assuming that the third- and higher order associations are zero, we can write the Bahadur model as follows:

$$P(\boldsymbol{Y}_i|\boldsymbol{X}_i, \boldsymbol{Z}_i) = \left\{ \prod_{j=1}^{n_i} \mu_{ij}^{Y_{ij}} (1 - \mu_{ij})^{1-Y_{ij}} \right\} \left(1 + \sum_{j<k} \rho_{i,jk} Y_{ij}^* Y_{ik}^* \right). \tag{5.28}$$

With this joint probability, we can use the law of total probability to derive the conditional distribution of $\boldsymbol{Y}_i^m|\boldsymbol{Y}_i^{\text{o}}, \boldsymbol{X}_i, \boldsymbol{Z}_i$. Note that both μ_{ij} and $\rho_{i,jk}$ are functions of parameter vector $\boldsymbol{\beta}$. We also write the joint probability as $P(\boldsymbol{Y}_i|\boldsymbol{X}_i, \boldsymbol{Z}_i; \boldsymbol{\beta})$.

For the missing covariate model, $\boldsymbol{X}_i^m|\boldsymbol{Y}_i^{\text{o}}, \boldsymbol{X}_i^{\text{o}}, \boldsymbol{Z}_i$, we consider a conditional model given the past observations. The joint density of $\boldsymbol{X}_i$ can be written as

$$P(\boldsymbol{X}_i|\boldsymbol{Y}_i^{\text{o}}, \boldsymbol{Z}_i) = P(X_{i1}|\boldsymbol{Z}_i) \prod_{j=2}^{n_i} P(X_{ij}|\bar{\boldsymbol{X}}_{ij}, \bar{\boldsymbol{Y}}_{ij}^{\text{o}}, \boldsymbol{Z}_i).$$

Again, $\bar{\boldsymbol{X}}_{ij}$ and $\bar{\boldsymbol{Y}}_{ij}^{\text{o}}$ represent the history of the covariate and outcome observed before time point j. To emphasize the nuisance parameter vector $\boldsymbol{\gamma}$ that specifies the missing covariate model, we also write the joint probability as $P(\boldsymbol{X}_i|\boldsymbol{Y}_i^{\text{o}}, \boldsymbol{Z}_i; \boldsymbol{\gamma})$. This model can be estimated by maximizing the observed likelihood function:

$$L_X(\gamma) = \prod_{i=1}^{N} \int P(\boldsymbol{X}_i|\boldsymbol{Y}_i^{\text{o}}, \boldsymbol{Z}_i) \mathrm{d}\boldsymbol{X}_i^m.$$

Now we turn to the augmented estimating equation and try to evaluate $\phi_{i,\text{opt}}$ in (5.27). When $\boldsymbol{X}_i^m$ is discrete, the augmented term becomes

$$\phi_{i,\text{opt}} = \sum_{\boldsymbol{X}_i^m, \boldsymbol{Y}_i^m} w_{ixy}(\boldsymbol{\beta}) \left\{ D_i' N_i (\boldsymbol{Y}_i - \boldsymbol{\mu}_i) \right\},$$

where

$$\begin{aligned} w_{ixy}(\boldsymbol{\beta}) &= P(\boldsymbol{Y}_i^m, \boldsymbol{X}_i^m | \boldsymbol{Y}_i^{\text{o}}, \boldsymbol{X}_i^{\text{o}}, \boldsymbol{Z}_i; \boldsymbol{\beta}, \widehat{\boldsymbol{\gamma}}) \\ &= P(\boldsymbol{Y}_i^m|\boldsymbol{Y}_i^{\text{o}}, \boldsymbol{X}_i, \boldsymbol{Z}_i; \boldsymbol{\beta}) P(\boldsymbol{X}_i^m|\boldsymbol{Y}_i^{\text{o}}, \boldsymbol{X}_i^{\text{o}}, \boldsymbol{Z}_i; \widehat{\boldsymbol{\gamma}}) \\ &= \frac{P(\boldsymbol{Y}_i|\boldsymbol{X}_i, \boldsymbol{Z}_i; \boldsymbol{\beta})}{\sum_{\boldsymbol{Y}_i^m} P(\boldsymbol{Y}_i|\boldsymbol{X}_i, \boldsymbol{Z}_i; \boldsymbol{\beta})} \cdot \frac{P(\boldsymbol{X}_i|\boldsymbol{Y}_i^{\text{o}}, \boldsymbol{Z}_i; \widehat{\boldsymbol{\gamma}})}{\sum_{\boldsymbol{X}_i^m} P(\boldsymbol{X}_i|\boldsymbol{Y}_i^{\text{o}}, \boldsymbol{Z}_i; \widehat{\boldsymbol{\gamma}})}. \end{aligned}$$

If $\boldsymbol{X}_i^m$ is continuous, the summation $\sum_{\boldsymbol{X}_i^m}$ is replaced by an integral $\int_{\boldsymbol{X}_i^m}$. A reject sampling procedure can be used to evaluate the integral. An EM type algorithm is proposed to solve equation (5.26) for $\boldsymbol{\beta}$, and the iteration procedures are described as follows:

1. Initialize the parameters $\boldsymbol{\beta}^{(0)}$.
2. In the kth iteration, $\boldsymbol{\beta}^{(k)}$ is obtained and used to update $w_{ixy}^{(k)}$, $\boldsymbol{M}_i^{(k)}$, and $\boldsymbol{N}_i^{(k)}$.
3. Treat $w_{ixy}^{(k)}$, $\boldsymbol{M}_i^{(k)}$, and $\boldsymbol{N}_i^{(k)}$ as fixed, and then solve the following equation:

$$\sum_{i=1}^{N} \left\{ \boldsymbol{D}_i' \boldsymbol{M}_i^{(k)} (\boldsymbol{Y}_i - \boldsymbol{\mu}_i) + \sum_{\boldsymbol{X}_i^m, \boldsymbol{Y}_i^m} w_{ixy}^{(k)} \left\{ \boldsymbol{D}_i' \boldsymbol{N}_i^{(k)} (\boldsymbol{Y}_i - \boldsymbol{\mu}_i) \right\} \right\} = 0,$$

   for $\boldsymbol{\beta}^{(k+1)}$.
4. Iterate steps 2 and 3 until convergence.

It was shown in Chen and Zhou (2011) that the estimator $\widehat{\boldsymbol{\beta}}$ is asymptotic normal and a sandwich variance formula is derived.

APPENDIX 5.A: TECHNICAL DETAILS OF THE APPROXIMATION METHODS FOR GLMM AND COMPUTER CODE FOR THE EXAMPLES

5.A.1 PQL and MQL

The PQL method borrows the idea of linearization used in the generalized linear models. Rewrite the mean structure model as

$$Y_{ij} = \mu_{ij}^{\mathrm{C}} + \varepsilon_{ij} = h\left(\boldsymbol{X}_{ij}'\boldsymbol{\beta} + \boldsymbol{Z}_{ij}'\boldsymbol{b}_i\right) + \varepsilon_{ij},$$

where h is the inverse link function g^{-1} and ε_{ij} is the error term with mean 0. Define ϕ to be the overdispersion parameter. With the canonical link function, the derivative of inverse link function, h', characterizes the mean–variance relationship:

$$V_{ij}^{\mathrm{C}} = \mathrm{var}\left(Y_{ij}|\boldsymbol{b}_i\right) = \phi h'[g(\mu_{ij}^{\mathrm{C}})] = \phi h'\left(\boldsymbol{X}_{ij}'\boldsymbol{\beta} + \boldsymbol{Z}_{ij}'\boldsymbol{b}_i\right).$$

By Taylor series expansion around the current $\widehat{\boldsymbol{\beta}}$ and $\widehat{\boldsymbol{b}}_i$, we have

$$\begin{aligned} Y_{ij} &\approx h(\boldsymbol{X}_{ij}'\widehat{\boldsymbol{\beta}} + \boldsymbol{Z}_{ij}'\widehat{\boldsymbol{b}}_i) + h'(\boldsymbol{X}_{ij}'\widehat{\boldsymbol{\beta}} + \boldsymbol{Z}_{ij}'\widehat{\boldsymbol{b}}_i)\boldsymbol{X}_{ij}'(\boldsymbol{\beta} - \widehat{\boldsymbol{\beta}}) + h'(\boldsymbol{X}_{ij}'\widehat{\boldsymbol{\beta}} \\ &\quad + \boldsymbol{Z}_{ij}'\widehat{\boldsymbol{b}}_i)\boldsymbol{Z}_{ij}'(\boldsymbol{b}_i - \widehat{\boldsymbol{b}}_i) + \varepsilon_{ij} \\ &\approx \hat{\mu}_{ij}^{\mathrm{C}} + \hat{V}_{ij}^{\mathrm{C}}\boldsymbol{X}_{ij}'(\boldsymbol{\beta} - \widehat{\boldsymbol{\beta}}) + \hat{V}_{ij}^{\mathrm{C}}\boldsymbol{Z}_{ij}'(\boldsymbol{b}_i - \widehat{\boldsymbol{b}}_i) + \varepsilon_{ij}, \end{aligned} \tag{5A.1}$$

where $\hat{\mu}_{ij}^{\mathrm{C}}$ and $\hat{V}_{ij}^{\mathrm{C}}$ are the current fit of *conditional* mean and variance, respectively. Define the working response variable

$$Y_{ij}^{*} = \frac{Y_{ij} - \hat{\mu}_{ij}^{\mathrm{C}}}{\hat{V}_{ij}^{\mathrm{C}}} + \boldsymbol{X}_{ij}^{\prime}\widehat{\boldsymbol{\beta}} + \boldsymbol{Z}_{ij}^{\prime}\widehat{\boldsymbol{b}}_{i}.$$

Reorganizing (5A.1) yields an LMM:

$$Y_{ij}^{*} \approx \boldsymbol{X}_{ij}^{\prime}\boldsymbol{\beta} + \boldsymbol{Z}_{ij}^{\prime}\boldsymbol{b}_{i} + \varepsilon_{ij}^{*}.$$

The new residual ε_{ij}^{*} is rescaled as $\varepsilon_{ij}^{*} = (\varepsilon_{ij} - \hat{\mu}_{ij}^{\mathrm{C}})/\hat{V}_{ij}^{\mathrm{C}}$. Therefore, one can iterate between fitting the LMM and updating Y_{ij}^{*} until convergence.

The MQL method is very similar to the PQL method, except that $h(\cdot)$ is expanded around $\widehat{\boldsymbol{\beta}}$ and $\boldsymbol{b}_i = 0$:

$$\begin{aligned} Y_{ij} &\approx h(\boldsymbol{X}_{ij}^{\prime}\widehat{\boldsymbol{\beta}}) + h^{\prime}(\boldsymbol{X}_{ij}^{\prime}\widehat{\boldsymbol{\beta}})\boldsymbol{X}_{ij}^{\prime}(\boldsymbol{\beta} - \widehat{\boldsymbol{\beta}}) + h^{\prime}(\boldsymbol{X}_{ij}^{\prime}\widehat{\boldsymbol{\beta}})\boldsymbol{Z}_{ij}^{\prime}\boldsymbol{b}_{i} + \varepsilon_{ij} \\ &\approx \hat{\mu}_{ij}^{\mathrm{M}} + \hat{V}_{ij}^{\mathrm{M}}\boldsymbol{X}_{ij}^{\prime}(\boldsymbol{\beta} - \widehat{\boldsymbol{\beta}}) + \hat{V}_{ij}^{\mathrm{M}}\boldsymbol{Z}_{ij}^{\prime}\boldsymbol{b}_{i} + \varepsilon_{ij}. \end{aligned} \tag{5A.2}$$

Here, $\hat{\mu}_{ij}^{\mathrm{M}}$ and $\hat{V}_{ij}^{\mathrm{M}}$ are the current fit of *marginal* mean and variance:

$$\begin{aligned} \hat{\mu}_{ij}^{\mathrm{M}} &= h(\boldsymbol{X}_{ij}^{\prime}\widehat{\boldsymbol{\beta}}), \\ V_{ij}^{\mathrm{M}} &= \phi h^{\prime}[g(\hat{\mu}_{ij}^{\mathrm{M}})] = \phi h^{\prime}(\boldsymbol{X}_{ij}^{\prime}\widehat{\boldsymbol{\beta}}). \end{aligned}$$

Hence, we have

$$Y_{ij}^{**} = \frac{Y_{ij} - \hat{\mu}_{ij}^{\mathrm{M}}}{\hat{V}_{ij}^{\mathrm{M}}} + \boldsymbol{X}_{ij}^{\prime}\widehat{\boldsymbol{\beta}} \approx \boldsymbol{X}_{ij}^{\prime}\boldsymbol{\beta} + \boldsymbol{Z}_{ij}^{\prime}\boldsymbol{b}_{i} + \varepsilon_{ij}^{**},$$

where the residual is rescaled as $\varepsilon_{ij}^{**} = (\varepsilon_{ij} - \hat{\mu}_{ij}^{\mathrm{M}})/\hat{V}_{ij}^{\mathrm{M}}$. The GLMM can be estimated by iterating between updating Y_{ij}^{**} and fitting the above LMM.

5.A.2 Laplace Approximation

The integrals in $L(\boldsymbol{\beta}, D)$ can be written in the form of $I = \int \exp(Q(\boldsymbol{b}))\mathrm{d}\boldsymbol{b}$. The idea of Laplace approximation is to approximate the integrand using a multivariate normal density, or equivalently, to replace $Q(\boldsymbol{b})$ by a quadratic form of $\boldsymbol{b}$. By the second-order Taylor series expansion of $Q(\boldsymbol{b})$ around its mode $\widehat{\boldsymbol{b}}$, we have

$$Q(\boldsymbol{b}) \approx Q(\widehat{\boldsymbol{b}}) + \frac{1}{2}(\boldsymbol{b} - \widehat{\boldsymbol{b}})^{\prime}Q^{\prime\prime}(\widehat{\boldsymbol{b}})(\boldsymbol{b} - \widehat{\boldsymbol{b}}),$$

and hence

$$I \propto \left|-Q''(\widehat{\boldsymbol{b}})\right|^{-1/2} \exp\left[Q(\widehat{\boldsymbol{b}})\right].$$

The mode $\widehat{\boldsymbol{b}}$ is solved from $Q'(\boldsymbol{b}) = 0$ as a function of $\boldsymbol{\beta}$ and D. Now the likelihood can be evaluated approximately in the closed form and could be maximized for estimating $\boldsymbol{\beta}$ and the variance components.

5.A.3 Gaussian Quadrature

The idea of the Gaussian quadrature is to use the bins to approximate the integral. The integral has the form $\int f(z)\phi(z)\mathrm{d}z$, where $\phi(z)$ is the density of multivariate normal distribution. It can be replaced by a weighted sum:

$$\int f(z)\phi(z)\mathrm{d}z \approx \sum_{q=1}^{Q} w_q f(z_q),$$

where z_q is the prespecified qth quadrature point or node (the position to use a bin to evaluate the integral) and w_q is the well-chosen weight. An algorithm is available to calculate all z_q and w_q for any number of quadrature points Q. The *adaptive Gaussian quadrature* is proposed by Pinheiro and Bates (1995) to reduce the approximation error. The idea is to rescale the nodes into the support of $f(z)\phi(z)$ and change the weights correspondingly. It is worth noting that the Laplace approximation is equivalent to the adaptive Gaussian quadrature with only one quadrature point.

5.A.4 Code for This Chapter

5.A.4.1 Stata Code Used in Section 5.5.1.1 The following code conducts the complete-case analysis for the IMPACT data:

Code

```
xtset sid
xi: xtmixed scl i.group*i.period || sid:period, ///
   vce(robust) cov(un)
```

5.A.4.2 Stata Code Used in Section 5.5.1.2 The following code is to conduct the available data analysis for the IMPACT study:

Code

```
xtset sid
xi: xtmixed scl i.group*i.period || sid:period, ///
   vce(robust) cov(un)
```

5.A.4.3 Stata Code Used in Section 5.5.1.3 The following Stata code illustrates the LOCF analysis for the IMPACT study:

Code

```
use impact_wide, clear
  keep if mtype<.

  # last value carry forward (LVCF)
  foreach x of numlist 3 6 12 18 24 {
      replace scl`x'=scl0 if mtype==5
  }
  foreach x of numlist 6 12 18 24 {
      replace scl`x'=scl3 if mtype==4
  }
  foreach x of numlist 12 18 24 {
      replace scl`x'=scl6 if mtype==3
  }
  foreach x of numlist 18 24 {
      replace scl`x'=scl12 if mtype==2
  }
  foreach x of numlist 24 {
      replace scl`x'=scl18 if mtype==1
  }

  keep sid nmiss mtype group scl*
  reshape long scl, i(sid) j(month)

  * linear mixed effects model
  xtset sid
  xi: xtmixed scl i.group*i.period || sid:period, ///
         vce(robust) cov(un)
```

5.A.4.4 Stata Code Used in Section 5.6.4 Here we present the Stata code for the IMPACT study using IPWGEE1 and IPWGEE2. In the following code, we estimate the probability that an outcome is observed at a specific time point, given the prior observed outcome variables and some selected baseline characteristics using logistic regression, and then generate both observation-level weights and cluster (subject)-level weights based on the estimated probabilities.

Code

```
use impact_wide, clear
  keep if mtype<.      // monotone missing only

  misstable pattern scl*, freq

  * indicators for being observed
  gen r0=1-dscl0
```

```
gen r3=1-dscl3
gen r6=1-dscl6
gen r12=1-dscl12
gen r18=1-dscl18
gen r24=1-dscl24

* predict conditional probability of being observed
    * when was also observed
* at previous time point
gen lam0=1  // pr(r1=1)
// conditional on observed at previous timepoint
xi: logistic r3 scl0 i.group i.site i.recmethd
predict lam3, pr

* by default, subjects dropped at 3 month were
    * excluded, predicted value missing
xi: logistic r6 scl0 scl3 i.group i.site ///
        i.recmethd
predict lam6, pr

xi: logistic r12 scl0 scl3 scl6 i.group ///
        i.site i.recmethd
predict lam12, pr

xi: logistic r18 scl0 scl3 scl6 scl12 i.group ///
        i.site i.recmethd
predict lam18, pr

xi: logistic r24 scl0 scl3 scl6 scl12 scl18 i.group ///
        i.site i.recmethd
predict lam24, pr

* observation level weights - probability of being
    * observed at each timepoint
gen owts0=1/lam0
gen owts3=1/(lam0*lam3)
gen owts6=1/(lam0*lam3*lam6)
gen owts12=1/(lam0*lam3*lam6*lam12)
gen owts18=1/(lam0*lam3*lam6*lam12*lam18)
gen owts24=1/(lam0*lam3*lam6*lam12*lam18*lam24)

* cluster(subject) level weights - probability of drop
    * out at observed dropout time
gen cwts=.
replace cwts=1/(lam0*lam3*lam6*lam12*lam18*lam24) ///
       if mtype==0
replace cwts=1/(lam0*lam3*lam6*lam12*lam18*(1-lam24))///
       if mtype==1
replace cwts=1/(lam0*lam3*lam6*lam12*(1-lam18))
```

```
        if mtype==2
    replace cwts=1/(lam0*lam3*lam6*(1-lam12)) if mtype==3
    replace cwts=1/(lam0*lam3*(1-lam6)) if mtype==4
    replace cwts=1/(lam0*(1-lam3)) if mtype==5
```

The following code implements IPWGEE1 using observation-level weights. Because this approach is not currently implemented in Stata, a workaround using `glm` command is implemented.

Code

```
    xi: glm scl i.group*i.period [pweight=owts], ///
        vce(cluster sid) family(gaussian) link(identity)
```

The following code then performs IPWGEE2 using cluster-level weights:

Code

```
  keep sid scl* group period* cwts mtype dscl*
    reshape long scl dscl period, i(sid) j(month)

  xtset sid month
    xi: xtgee scl i.group*i.period [pweight=cwts], ///
        vce(robust) corr(uns)
```

5.A.4.5 R Code Used in Section 5.8.3 Here, we present the R code for implementing multiple imputation analysis for the NACC UDS data. The package `mice` is used. The following code appends the transformation of variables, dummy variables, and interaction terms.

```
> attach(NACCUDS)
> NACCUDS2=cbind(NACCUDS,cons=1,logcdr=log(cdr+1),
+ mmse2=logit((mmse+0.5)/31),edu2=(education==2)+0,
+ edu3=(education==3)+0,hachin2=(hachin==2)+0,
+ hachin3=(hachin==3)+0,age.year=age0*year,
+ gender.year=gender*year,edu.year2=(education==2)*year,
+ edu.year3=(education==3)*year,his.year2=(hachin==2)*year,
+ his.year3=(hachin==3)*year,ap.year=apoee4*year,
+ mmse.year=mmse*year)
> detach()
```

A ‘‘dry run’’ with the maximum number of iteration set to 0 is the first step to initialize the imputation parameters.

```
> ini=mice(NACCUDS2,maxit=0)
```

Then the imputation methods are specified.

```
> meth=ini$meth
> meth["mmse2"]="2l.norm"
> meth["mmse"]="~I(round(1/(1+exp(-mmse2))*31-0.5))"
> meth["hachin"]="pmm"
> meth["education"]="2lonly.pmm"
> meth["apoee4"]="2lonly.pmm"
> meth["edu2"]="~I(education==2)"
> meth["edu3"]="~I(education==3)"
> meth["hachin2"]="~I(hachin==2)"
> meth["hachin3"]="~I(hachin==3)"
> meth["edu.year2"]="~I((education==2)*year)"
> meth["edu.year3"]="~I((education==3)*year)"
> meth["his.year2"]="~I((hachin==2)*year)"
> meth["his.year3"]="~I((hachin==3)*year)"
> meth["ap.year"]="~I(apoee4*year)"
> meth["mmse.year"]="~I(mmse*year)"
```

The next step is to specify the prediction matrix of the imputation model. The following is a print of our specified prediction matrix:

```
> pred[c("mmse2","education","apoee4","hachin"),]
          mmse age0 gender education apoee4 year cdr hachin ID
mmse2        0    1      1         0      1    1   0      0 -2
education    1    1      1         0      1    0   0      0 -2
apoee4       1    1      1         0      0    0   0      0 -2
hachin       1    1      1         0      1    1   0      0  0
          cons logcdr mmse2 edu2 edu3 hachin2 hachin3 age.year
mmse2        2      1     0    1    1       1       1        1
education    0      1     0    0    0       1       1        0
apoee4       0      1     0    1    1       1       1        0
hachin       0      1     0    1    1       0       0        1
          gender.year edu.year2 edu.year3 his.year2 his.year3
mmse2               1         1         1         1         1
education           0         0         0         0         0
apoee4              0         0         0         0         0
hachin              0         1         1         0         0
           ap.year mmse.year
mmse2            1         0
education        0         0
apoee4           0         0
hachin           1         1
```

The visit scheme specifies the order of imputation. We need to keep the passive imputations synchronized with the imputed variables.

```
> vis=ini$vis
> vis["mmse2"]=1
> vis["mmse"]=12
> vis
     mmse education     apoee4      hachin       mmse2        edu2
       12         4          5           8           1          13
     edu3   hachin2    hachin3 edu.year2 edu.year3 his.year2
       14        15         16          19          20          21
his.year3   ap.year  mmse.year
       22        23         24
```

We also try imputing MMSE using the predictive mean matching method. The `pmm` does not make any distributional assumption, but only requires a mean model with identity link.

```
> meth2=meth
> meth2["mmse2"]="pmm"
> pred2=pred
> pred2["mmse2","ID"]=0
> pred2["mmse2","cons"]=0
```

The following code generates 20 complete data sets with missing values imputed by the two methods:

```
> imp2=mice(NACCUDS2,m=20,me=meth2,pred=pred2,vis=vis,maxit=20,
+        seed=10065)
> imp1=mice(NACCUDS2,m=20,me=meth,pred=pred,vis=vis,maxit=20,
+        seed=7523)
> plot(imp1,c("hachin","education","apoee4","mmse"),
+    layout=c(2,4))
```

The analysis for the 20 imputed data sets is conducted by the following code.

```
> library(gee)
> fitC=gee(mmse~(I((age0-75)/10)+gender+apoee4+
+        as.factor(education)+as.factor(hachin))*year,
+        corstr="independence",id=ID,data=NACCUDS)
> fit1=with(imp1,gee(mmse~(I((age0-75)/10)+gender+apoee4+
+         as.factor(education)+as.factor(hachin))*year,
+         corstr="independence",id=ID))
> fit2=with(imp2,gee(mmse~(I((age0-75)/10)+gender+apoee4+
+         as.factor(education)+as.factor(hachin))*year,
+         corstr="independence",id=ID))
```

The code for combining fit1 using Rubin's rule is given below.

```
> PE1=VAR1=matrix(0,20,16)
> for(i in 1:20){
+ PE1[i,]=(fit1$analyses[[i]])$coef
+ VAR1[i,]=summary(fit1$analyses[[i]])$coef[,4]^2
+ }
> COEF1=apply(PE1,2,mean)
> U1=apply(VAR1,2,mean)
> B1=apply(PE1,2,var)
> SE1=(U1+(1+1/20)*B1)^.5
> RES1=cbind(COEF1,SE1)
```

5.A.4.6 Stata Code Used in Section 5.8.4.1 The following Stata code is for the analysis of multiple imputation with monotone missingness:

Code

```
//==> Monotone imputation: recommended for true monotone
//==> missingness
    use impact_wide, clear
    keep if mtype<.

    * keep the data in wide format, so one row is records
        * for one subject
        * 1) mi set your data
    mi set wide

    * 2) Use mi describe often
    mi describe

    * 3) identify missing values
    mi misstable summarize

    * 4) register variables we wish to impute
    mi register imputed scl*
    mi describe

    * 5) impute the missing outcomes. we first use monotone
        *    to take into account
    * within-subject clustering.  Note later scl values
        * will be imputed using the imputed earlier scl
    values set seed 45078

    * didn't use satis00 and inc400 due to too many missing
        * values
    * stratified by group, i.e., do imputation separately
        * for groups
    mi impute monotone (regress) scl0 scl3 scl6 scl12 ///
            scl18 scl24 = site recmethd age male white ///
```

```
        married educat work00 pref00 ghlth00 numdis1, ///
        by(group) force add(10)

    * reshape the data to long format
mi reshape long scl period, i(sid) j(month)

* mixed effects model
mi xtset sid
xi: mi estimate: xtmixed scl i.group*i.period || ///
        sid:period, vce(robust) cov(un)
```

5.A.4.7 *Stata Code Used in Section 5.8.4.2* The following is the Stata code for multiple imputation based on the multivariate normal distribution for the IMPACT data:

Code

```
set seed 45078
mi impute mvn scl0 scl3 scl6 scl12 scl18 scl24 = ///
   site recmethd age male white married educat work00 ///
   pref00 ghlth00 numdis1, by(group) force add(10)
```

5.A.4.8 *Stata Code Used in Section 5.8.4.3* The multiple imputation with chained equations is applied to the IMPACT data and achieved by the following chunk of code:

Code

```
set seed 45078
mi impute chained (regress) scl0 scl3 scl6 scl12 scl18 ///
   scl24 = site recmethd age male white married educat ///
   work00 pref00 ghlth00 numdis1, by(group) force add(10)
```

5.A.4.9 *Stata Code Used in Section 5.8.4.4* The following is the Stata code to conduct analysis of IMPACT data based on the multiple imputation with chained equations for a general missing data pattern. We recommend readers to start by following the steps listed in the following code chunk. One tip is to use ''mi describe'' often so that we can keep a close eye on our imputation.

Code

```
//==> mice for general missing data pattern
  use impact_wide, clear

    * 1) set how imputed data are stored
    mi set wide

    * 2) check missing data pattern
    mi describe
    mi misstable summarize
```

```
* 3) let stata know which variable you want to impute
mi register imputed white pref00 numdis1 ghlth00 ///
      scl* inc400
mi describe

* 4) imputation with chained equations
set seed 45078
mi impute chained ///
      (logit) white (ologit) pref00 ghlth00 (poisson) ///
      numdis1 (pmm) inc400 ///
      (regress) scl0 scl3 scl6 scl12 scl18 scl24 = ///
      site recmethd age male married educat work00, ///
      by(group) force add(10)

* 5) run your analysis
mi reshape long scl period, i(sid) j(month)
mi xtset sid
xi: mi estimate: xtmixed scl i.group*i.period || ///
      sid:period, vce(robust) cov(un)
```

CHAPTER 6

SURVIVAL ANALYSIS UNDER IGNORABLE MISSINGNESS

6.1 OVERVIEW

We introduced the cross-sectional and longitudinal studies in Chapters 4 and 5, where the outcomes are completely observed or missing. In survival analysis, the outcome of interest, the failure time, may be censored such that we may observe either the failure time or the censoring time (see the Pathways example in Section 1.3.6). This makes this chapter different from Chapters 4 and 5. One of the basic tasks of survival analysis is to model the relationship between failure time and covariates. Due to the expensive examination fee or other reasons, some of the covariates might not be observed. In this chapter, we provide methods for survival analysis when the covariates are possibly missing. We mainly focus on the Cox proportional hazards model because it is the most popular model in survival analysis and most of the literature on survival analysis with missing covariates focused on it. We also discuss the semiparametric transformation regression model, a more general class that includes the Cox proportional hazards model as a special case.

This chapter is organized as follows. In Section 6.2, we introduce some commonly used notations and review the Cox regression when covariate data are completely observed. Then several methods are offered. The enhanced complete-case analysis is introduced in Section 6.3, and the weighted methods are given in Section 6.4. Imputation methods are provided in Section 6.5. The nonparametric maximum likelihood (NPML) estimation is presented in Section 6.6 and its application to the

Applied Missing Data Analysis in Health Sciences, First Edition. Xiao-Hua Zhou, Chuan Zhou, Danping Liu, and Xiaobo Ding.

transformation models is given in Section 6.7. The methods are illustrated in Section 6.8 using the Pathways study as an example. Concluding remarks are given in Section 6.9.

6.2 INTRODUCTION

Before we proceed to discuss the methods handling the missing-covariate problem, we would like to give some introductions. The Cox proportional hazards model without any missing values is reviewed in Section 6.2.1. This helps us understand the problems that arise in case of missing covariates and how to deal with them. The missing mechanisms are introduced in Section 6.2.2.

6.2.1 Review of the Cox Model with Completely Observed Covariates

In survival analysis, the Cox proportional hazards model is most widely used. Proposed by Cox (1972), the model has been studied by many authors, making the original paper one of the most frequently cited journal articles in statistics and medicine. Here, we give a brief review on it when the covariates are completely observed.

Suppose we have a sample of size n. Let T_i be the failure time of the ith subject. In the usual scenarios of survival analysis, T_i might be censored. Let C_i be the censoring time. The observed time is $X_i = \min(T_i, C_i)$. A censoring indicator, $\Delta_i = I(T_i \leq C_i)$, is introduced. That is, $\Delta_i = 1$ when T_i is observed, and 0 when it is censored. Let $N_i(t) = \Delta_i I\,(X \leq t)$ and $Y_i(t) = I(X_i \geq t)$ be the counting and at-risk processes, respectively. A column vector of covariates is denoted by $\mathbf{Z}_i$. If the covariates are time dependent, then the vector of covariate values at time t is denoted by $\mathbf{Z}_i(t)$. In this chapter, we mainly discuss the situations where $\mathbf{Z}_i$ is time independent. However, we also discuss which methods are adaptive for the time-dependent cases.

In the Cox model, the hazard function at time t for the ith individual ($i = 1, \dots, n$) is assumed to be

$$\lambda(t|\mathbf{Z}_i) = \lambda_0(t)\exp(\beta^\top \mathbf{Z}_i),$$

where β is the regression coefficient vector and $\lambda_0(t)$ is the baseline hazard function, which is completely unknown. Usually the estimation of β is of interest for its interpretation of effects of covariates. The commonly used estimation method is the partial likelihood proposed by Cox (1975).

Define

$$\mathbf{s}^{(k)}(\beta, t) = \mathbb{E}\left\{Y(t)\exp\left(\beta^\top \mathbf{Z}\right)\mathbf{Z}^{\otimes k}\right\}, \quad \text{for } k = 0, 1, 2,$$

$$\mathbf{e}(\beta, t) = \mathbf{s}^{(1)}(\beta, t)\,/\,s^{(0)}(\beta, t),$$

$$\mathbf{v}(\beta, t) = \frac{\mathbf{s}^{(2)}(\beta, t)}{s^{(0)}(\beta, t)} - \mathbf{e}(\beta, t)^{\otimes 2},$$

where for a column vector $\mathbf{a}$, $\mathbf{a}^{\otimes 0} = 1$, $\mathbf{a}^{\otimes 1} = \mathbf{a}$, and $\mathbf{a}^{\otimes 2} = \mathbf{a}\mathbf{a}^\top$. Here $\mathbb{E}$ denotes expectation. Their corresponding estimators are

$$\mathbf{S}^{(k)}(\beta, t) = n^{-1} \sum_{i=1}^{n} Y_j(t) \exp\left(\beta^\top \mathbf{Z}_j\right) \mathbf{Z}_j^{\otimes k}, \quad \text{for } k = 0, 1, 2,$$

$$\mathbf{E}(\beta, t) = \mathbf{S}^{(1)}(\beta, t) / S^{(0)}(\beta, t),$$

and

$$\mathbf{V}(\beta, t) = \frac{\mathbf{S}^{(2)}(\beta, t)}{S^{(0)}(\beta, t)} - \mathbf{E}(\beta, t)^{\otimes 2},$$

respectively.

If subjects are completely observed, the partial likelihood score function is

$$\mathbf{U}_{\mathrm{f}}(\beta) = \sum_{i=1}^{n} \Delta_i \left\{\mathbf{Z}_i - \mathbf{E}\left(\beta, X_i\right)\right\}. \tag{6.1}$$

Note that $\mathbf{E}\left(\beta, X_i\right)$ is the conditional expectation of $\mathbf{Z}_i$ on all prior information up to time X_i. Solving the estimating equation $\mathbf{U}_{\mathrm{f}}(\beta) = 0$ using the Newton–Raphson algorithm, we can obtain the maximum partial likelihood estimator of β, say $\hat{\beta}$. Begun et al. (1983) showed that it is semiparametrically efficient. Let

$$M(t) = N(t) - \int_0^\tau Y(u) \exp\left(\beta^\top \mathbf{Z}\right) \lambda_0(u)\, \mathrm{d}u \tag{6.2}$$

be the counting process martingale for the failure process, and

$$\mathbf{M}_{\tilde{Z}} = \int_0^\tau [\mathbf{Z} - \mathbf{e}(\beta, t)]\, \mathrm{d}M(t) \tag{6.3}$$

the martingale transformation with mean 0 and variance $\Sigma = \mathbb{E}\left\{\mathbf{M}_{\tilde{Z}}^{\otimes 2}\right\}$. Andersen and Gill (1982) and Tsiatis (1981) showed that the maximum partial likelihood estimator of β is asymptotically normal, that is,

$$n^{1/2}\left(\hat{\beta} - \beta\right) \xrightarrow{d} N\left(0, \Sigma^{-1}\right), \quad \text{as } n \to \infty.$$

The limiting variance of $\hat{\beta}$ can be obtained by calculating

$$\hat{\Sigma} = -n^{-1} \frac{\partial \mathbf{U}_{\mathrm{f}}\left(\hat{\beta}\right)}{\partial \beta} = n^{-1} \sum_{i=1}^{n} \Delta_i \mathbf{V}\left(\hat{\beta}, X_i\right).$$

6.2.2 Missing Data

Due to survey nonresponse, high cost, or other reasons, some of the covariates might be missing. Without loss of generality, we assume that the covariate vector can be partitioned as $\mathbf{Z}_i = \left(\mathbf{Z}_{1i}^{\top}, \mathbf{Z}_{2i}^{\top}\right)^{\top}$, where $\mathbf{Z}_{1i}$ is a completely observed vector of covariates and $\mathbf{Z}_{2i}$ can be missing. The dimensions of $\mathbf{Z}_{1i}$, $\mathbf{Z}_{2i}$, and $\mathbf{Z}_i$ are denoted by p, q, and $r = p + q$, respectively. The component-wise missing indicator is denoted by R_{ji} for $j = 1, \ldots, r$. That is, $R_{ji} = 1$ when the jth component of $\mathbf{Z}_i$ is observed, and 0 otherwise. Of course, $R_{1i}, \ldots, R_{pi} \equiv 1$. Let $R_{0i} = \prod_{j=1}^{r} R_{ji}$ denote the overall missing indicator, that is, $R_{0i} = 1$ when all the components of $\mathbf{Z}_i$ are observed, and 0 otherwise. We also let $\mathbf{R}_i$ be a $r \times r$ diagonal matrix with indicator functions $\{R_{1i}, \ldots, R_{ri}\}$ as the diagonal elements. Accordingly,

$$\mathbf{R}_i = \begin{pmatrix} \mathbf{I}_p & \mathbf{0}_{p\times q} \\ \mathbf{0}_{q\times p} & \mathbf{R}_{2i} \end{pmatrix},$$

where $\mathbf{I}_p$ is an identity matrix with dimension $p \times p$, $\mathbf{0}_{p\times q}$ is a zero matrix with dimension $p \times q$, and $\mathbf{R}_{2i}$ is a $q \times q$ diagonal matrix with indicator functions $\{R_{p+1,i}, \ldots, R_{ri}\}$ as the diagonal elements. A special case of missingness is that all the components of $\mathbf{Z}_i$ are possibly missing, that is, $p = 0$ and $q = r$.

It is commonly assumed that data are missing completely at random (MCAR) or missing at random (MAR). The MCAR assumption is that

$$R_{ji} \perp\!\!\!\perp \left\{\mathbf{Z}_{1i}, \mathbf{Z}_{2i}, X_i, \Delta_i\right\}, \quad \text{for } j = 1, \ldots, r,$$

where "$\perp\!\!\!\perp$" represents the statistical independence. The above formula means that missingness probability does not depend on any variables. The MAR assumption is that

$$R_{ji} \perp\!\!\!\perp \mathbf{Z}_{2i} | \mathbf{W}_i, \tag{6.4}$$

where $\mathbf{W}_i$ is a set of the observed values of the subject i. In some practical situations, $\mathbf{W}_i = \mathbf{Z}_{1i}$. For example, in a clinical trial, treatment assignment and some variables may be completely observed, but the probability of observing the other variables may be determined only by the treatment assignment. Usually, $\mathbf{W}_i = \left(X_i, \Delta_i, \mathbf{Z}_{1i}^{\top}\right)^{\top}$. For example, some covariates are more likely to be missing for the subjects who are more sick and, therefore, may have shorter failure times.

From (6.1) it can be seen that if one uses the partial likelihood method to deal with the missing covariate problem, two issues should be taken into account. Note that $\mathbf{E}(\beta, t)$ in (6.1) is an estimate of $\mathbf{e}(\beta, t)$ when data are completely observed. The first issue is to replace $\mathbf{E}(\beta, t)$ with an estimate of $\mathbf{e}(\beta, t)$ in the presence of missing covariates. Once the estimate of $\mathbf{e}(\beta, t)$ is available, we can see that each component of $\mathbf{Z}_i$ contributes to each component of the partial likelihood score. This is a special structure of the partial likelihood score. So the second issue is how to deal with the subjects with missing values in the special structure.

It is well known that the complete-case analysis and inverse probability weighted methods are popular ways of dealing with missing data problems, which can also be applied to Cox regression with missing covariates. Due to the special structure of the partial likelihood score function, it is interesting to see that these methods can be more sophisticated. In Section 6.3, we introduce the enhanced complete-case analysis, which improves the complete-case analysis under the MCAR assumption. The weighted methods are introduced in Section 6.4.

6.3 ENHANCED COMPLETE-CASE ANALYSIS

The complete-case analysis is a naive and simple way to deal with missing data, since it excludes all the subjects with missing values and treats the remaining subjects as a full data set. For the complete-case analysis, $\mathbf{s}^{(k)}(\beta, t)$ can be estimated by

$$\mathbf{S}_{\mathrm{cc}}^{(k)}(\beta, t) = n^{-1} \sum_{j=1}^{n} R_{0j} Y_j(t) \exp\left(\beta^\top \mathbf{Z}_j\right) \mathbf{Z}_j^{\otimes k}, \quad \text{for } k = 0, 1, 2,$$

and $\mathbf{e}(\beta, t)$ is estimated by

$$\mathbf{E}_{\mathrm{cc}}(\beta, t) = \mathbf{S}_{\mathrm{cc}}^{(1)}(\beta, t) / S_{\mathrm{cc}}^{(0)}(\beta, t).$$

Then, the complete-case partial likelihood score is

$$\mathbf{U}_{\mathrm{cc}}(\beta) = \sum_{i=1}^{n} \Delta_i R_{0i} \left\{ \begin{pmatrix} \mathbf{Z}_{1i} \\ \mathbf{Z}_{2i} \end{pmatrix} - \mathbf{E}_{\mathrm{cc}}(\beta, X_i) \right\}.$$

Typically, the complete-case estimator is consistent only when $\mathbf{Z}_{2i}$ are missing completely at random.

It can be seen from $\mathbf{U}_{\mathrm{cc}}(\beta)$ that a subject i with missing $\mathbf{Z}_{2i}$ makes no contributions to $\mathbf{U}_{\mathrm{cc}}$. However, the observed values of the ith subject can still make contributions to the corresponding components of the score function, which can be used to increase efficiency of statistical inference. So a more efficient estimator, called enhanced complete-case estimator, $\hat{\beta}_{\mathrm{ecc}}$, can be obtained by solving the following score function:

$$\begin{aligned}
\mathbf{U}_{\mathrm{ecc}}(\beta) &= \sum_{i=1}^{n} \Delta_i \mathbf{R}_i \left\{ \mathbf{Z}_i - \mathbf{E}_{\mathrm{cc}}(\beta, X_i) \right\} \\
&= \sum_{i=1}^{n} \Delta_i \left\{ \begin{pmatrix} \mathbf{Z}_{1i} \\ \mathbf{R}_{2i}\mathbf{Z}_{2i} \end{pmatrix} - \mathbf{E}_{\mathrm{cc}}(\beta, X_i) \right\}.
\end{aligned}$$

This estimator was proposed by Lin and Ying (1993). They called it approximate partial likelihood estimator. But we would rather call it the enhanced complete-case analysis because it is derived from the complete-case analysis, and it is consistent only when the missing mechanism is missing completely at random as well as the complete-case analysis. The only difference between $\mathbf{U}_{cc}$ and $\mathbf{U}_{ecc}$ is that R_{0i} in $\mathbf{U}_{cc}$ is replaced by $\mathbf{R}_i$ in $\mathbf{U}_{ecc}$. As Lin and Ying (1993) indicated, the enhanced complete-case estimator is generally more efficient than the complete-case estimator. The gain in efficiency will be most profound, if the covariates of primary interest (e.g., treatment indicators) have measurements on (virtually) all subjects but the secondary covariates are heavily missing.

The enhanced complete-case estimator can be shown to be consistent and asymptotically normal at rate root n under some regular conditions. In addition, a simple and consistent estimator for the limiting covariance matrix can be easily constructed. From theoretical considerations, the limiting covariance matrix of the enhanced complete-case estimator is $\mathbf{A}^{-1}(\beta)\mathbf{B}(\beta)\left[\mathbf{A}^{-1}(\beta)\right]^{\top}$ at the true parameter β. Here, $\mathbf{B}(\beta) = \mathbb{E}\left\{\mathbf{W}_1(\beta)^{\otimes 2}\right\}$ and

$$\mathbf{W}_i(\beta) = \Delta_i \mathbf{R}_i \left\{\mathbf{Z}_i - \mathbf{e}\left(\beta, X_i\right)\right\} - \int_0^{X_i} \frac{\mathbf{r}}{r_0} R_{0i} \exp\left(\beta^{\top}\mathbf{Z}_i\right)\left\{\mathbf{Z}_i - \mathbf{e}(\beta, t)\right\} \lambda_0(t)\,\mathrm{d}t,$$

where $\mathbf{r} = \mathbb{E}\left(\mathbf{R}_i\right)$ and $r_0 = \mathbb{E}\left(R_{0i}\right)$. We can estimate $\mathbf{B}(\beta)$ by $\mathbf{B}_n(\beta) = n^{-1}\sum_{i=1}^n \tilde{\mathbf{W}}_i(\beta)^{\otimes 2}$, where

$$\begin{aligned}\tilde{\mathbf{W}}_i(\beta) = \Delta_i \mathbf{R}_i & \left\{\mathbf{Z}_i - \mathbf{E}\left(\beta, X_i\right)\right\} \\ & -\frac{1}{n}\sum_{l=1}^{n} \Delta_l Y_i\left(X_l\right) R_{0i}\mathbf{R}_l \exp\left(\beta^{\top}\mathbf{Z}_i\right) \\ & \qquad \times\left\{\mathbf{Z}_i - \mathbf{E}_{cc}\left(\beta, X_l\right)\right\} / S^{(0)}\left(\beta, X_l\right).\end{aligned}$$

In addition, $\mathbf{A}(\beta) = \lim_{n\to\infty}\mathbf{A}_n(\beta)$ and $\mathbf{A}_n(\beta) = -n^{-1}\partial\mathbf{U}_{ecc}(\beta)/\partial\beta$, where

$$\partial\mathbf{U}_{ecc}(\beta)/\partial\beta = -\sum_{i=1}^{n}\Delta_i\mathbf{R}_i\left\{\mathbf{S}_{cc}^{(2)}\left(\beta, X_i\right)/S_{cc}^{(0)}\left(\beta, X_i\right) - \mathbf{E}_{cc}\left(\beta, X_i\right)^{\otimes 2}\right\}.$$

Thus, we can obtain that a consistent estimator of the limiting covariance matrix is $A_n^{-1}\left(\hat{\beta}_{ecc}\right)B_n\left(\hat{\beta}_{ecc}\right)\left[A_n^{-1}\left(\hat{\beta}_{ecc}\right)\right]^{\top}$.

It is worth noting that the complete-case and the enhanced complete-case estimations are adaptive in the cases where all components of $\mathbf{Z}$ are possibly missing, that is, $\mathbf{Z}_1$ is null and $p = 0$. In fact, in these cases, we can define a more general missing mechanism. Among the methods that will be introduced later, the NPML method can deal with this missing mechanism, which will be defined there in detail. Another advantage of the complete-case and enhanced complete-case analysis is that they are adaptive for the time-dependent covariates.

6.4 WEIGHTED METHODS

The weighted methods play an important role in dealing with missing data. One very early weighted estimator is the Horvitz–Thompson estimator proposed by Horvitz and Thompson (1952). Then Robins et al. (1994) proposed a more efficient and robust estimator than the Horvitz–Thompson estimator. In this section, we introduce the simple weighted and fully augmented weighted methods, and their corresponding reweighting methods for the Cox regression model.

6.4.1 Simple Weighted Estimations

In this section, we introduce the simple weighted estimations of β. Denote the selection probability by π, that is, $\pi_i = P\left(R_{0i} = 1 | X_i, \Delta_i, \mathbf{Z}_i\right)$. Suppose that the MAR assumption is satisfied. Then, we have $\pi_i = P\left(R_{0i} = 1 | \mathbf{W}_i\right)$, where $\mathbf{W}_i$ contains some or all of the observed variables. Generally, we could let $\mathbf{W}_i = \left\{X_i, \Delta_i, \mathbf{Z}_{1i}\right\}$.

We first consider the case where the selection probability π is known. When π is unknown, we can impose a parametric model for it. Once the estimator of π is obtained, the simple weighted estimation is carried out as if π was known. It is worth noting that the resulting variances of the estimators of β when π is known or estimated are different. We cannot ensure that the imposed parametric model for π is correctly specified. The incorrect specification of π may result in a large bias of estimation of β. So, nonparametric estimation of π is also considered.

6.4.1.1 When π Is Known Suppose the selection probabilities π_i are known. The simple weighted estimators of $\mathbf{s}^{(k)}(\beta, t)$ and $\mathbf{e}(\beta, t)$ are

$$\mathbf{S}_{\mathrm{sw}}^{(k)}(\beta, \pi, t) = n^{-1} \sum_{j=1}^{n} \left(R_{0j}/\pi_j\right) Y_j(t) \exp\left(\beta^\top \mathbf{Z}_j\right) \mathbf{Z}_j^{\otimes k}$$

and

$$\mathbf{E}_{\mathrm{sw}}(\beta, \pi, t) = \mathbf{S}_{\mathrm{sw}}^{(1)}(\beta, \pi, t) / S_{\mathrm{sw}}^{(0)}(\beta, \pi, t),$$

respectively. Then, the simple weighted score function is defined as

$$\mathbf{U}_{\mathrm{sw}}(\beta, \pi) = \sum_{i=1}^{n} \frac{R_{0i}}{\pi_i} \Delta_i \left\{\mathbf{Z}_i - \mathbf{E}_{\mathrm{sw}}\left(\beta, \pi, X_i\right)\right\}. \tag{6.5}$$

By solving the equation $\mathbf{U}_{\mathrm{sw}}(\beta, \pi) = 0$, we obtain the estimator of β, which is denoted by $\hat{\beta}_{\mathrm{sw}}(\pi)$.

Recall that the counting process martingale for the failure process $M(t)$ and the martingale transformation $\mathbf{M}_{\tilde{Z}}$ are defined in (6.2) and (6.3), respectively. It is shown that under some regular conditions,

$$\sqrt{n}\left(\hat{\beta}_{\mathrm{sw}}(\pi)-\beta\right)\xrightarrow{d} N\left(0, \Sigma^{-1}\Sigma_{\mathrm{sw}}(\pi)\Sigma^{-1}\right),$$

where $\Sigma=\mathbb{E}\left\{\mathbf{M}_{\tilde{Z}}^{\otimes 2}\right\}$ and $\Sigma_{\mathrm{sw}}(\pi)=\mathbb{E}\left\{\pi^{-1}\mathbf{M}_{\tilde{Z}}^{\otimes 2}\right\}=\Sigma+\mathbb{E}\left\{\left(\pi^{-1}-1\right)\mathbf{M}_{\tilde{Z}}^{\otimes 2}\right\}$. Note that the first term of $\Sigma_{\mathrm{sw}}(\pi)$, Σ, pertains to the Cox partial likelihood estimator based on full cohort data, whereas the second term quantifies the efficiency loss due to missing covariates.

The limiting variance matrix of $\hat{\beta}_{\mathrm{sw}}(\pi)$ can be estimated by a plug-in method; that is, it is obtained by estimating Σ and $\Sigma_{\mathrm{sw}}(\pi)$. Let

$$\mathbf{V}_{\mathrm{sw}}(\beta,\pi,t)=\frac{\mathbf{S}_{\mathrm{sw}}^{(2)}(\beta,\pi,t)}{S_{\mathrm{sw}}^{(0)}(\beta,\pi,t)}-\mathbf{E}_{\mathrm{sw}}(\beta,\pi,t)^{\otimes 2}.$$

It can be easily obtained that Σ can be estimated consistently by

$$\hat{\Sigma}=\sum_{i=1}^{n}\Delta_i\frac{R_{0i}}{\pi_i}\mathbf{V}_{\mathrm{sw}}\left(\hat{\beta}_{\mathrm{sw}}(\pi),\pi,X_i\right). \tag{6.6}$$

To estimate $\Sigma_{\mathrm{sw}}(\pi)$, we first obtain

$$\mathrm{d}\hat{\Lambda}_0^{\mathrm{sw}}(t,\pi)=\frac{1}{n}\frac{\sum_{j=1}^{n}\left(R_{0j}/\pi_j\right)\mathrm{d}N_j(t)}{S_{\mathrm{sw}}^{(0)}\left(\hat{\beta}_{\mathrm{sw}}(\pi),\pi,t\right)},$$

$$\mathrm{d}\hat{M}(t)=dN_i(t)-Y_i(t)\exp\left\{\hat{\beta}_{\mathrm{sw}}^{\mathsf{T}}(\pi)\mathbf{Z}_i\right\}\mathrm{d}\hat{\Lambda}_0^{\mathrm{sw}}(t,\pi),$$

and let

$$\hat{\mathbf{M}}_{\tilde{Z},i}=\int_0^{\tau}\left[\mathbf{Z}_i-\mathbf{E}_{\mathrm{sw}}\left(\hat{\beta}_{\mathrm{sw}}(\pi),\pi,t\right)\right]\mathrm{d}\hat{M}(t), \tag{6.7}$$

which can be further rewritten as

$$\begin{aligned}\hat{\mathbf{M}}_{\tilde{Z},i}&=\Delta_i\left\{\mathbf{Z}_i-\mathbf{E}_{\mathrm{sw}}\left(\hat{\beta}_{\mathrm{sw}}(\pi),\pi,X_i\right)\right\}\\&\quad-\exp\left\{\hat{\beta}_{\mathrm{sw}}^{\mathsf{T}}(\pi)\mathbf{Z}_i\right\}\sum_{j=1}^{n}\frac{R_{0j}\Delta_j Y_i\left(X_j\right)\left\{\mathbf{Z}_i-\mathbf{E}_{\mathrm{sw}}\left(\hat{\beta}_{\mathrm{sw}}(\pi),\pi,X_j\right)\right\}}{n\pi_j S_{\mathrm{sw}}^{(0)}\left(\hat{\beta}_{\mathrm{sw}}(\pi),\pi,X_j\right)}.\end{aligned}$$

Then, $\Sigma_{\mathrm{sw}}(\pi)$ can be estimated consistently by

$$\hat{\Sigma}_{\mathrm{sw}}(\pi) = n^{-1} \sum_{i=1}^{n} R_{0i} \pi_i^{-2} \hat{\mathbf{M}}_{\tilde{Z},i}^{\otimes 2}.$$

So the estimator of the limiting variance of $\hat{\beta}_{\mathrm{sw}}(\pi)$ is $\hat{\Sigma}^{-1}\hat{\Sigma}_{\mathrm{sw}}(\pi)\hat{\Sigma}^{-1}$.

6.4.1.2 *Model π Parametrically* In most cases, π is unknown and needs to be estimated. We can model π parametrically or nonparametrically. Pugh et al. (1994) and Xu et al. (2009) used a parametric model to characterize the selection probability, that is, π is known up to a finite number of parameters. Specifically, let $\pi(\mathbf{W}) = \pi(\mathbf{W}; \alpha)$, where α is a finite-dimensional parameter vector. For example, under the common logistic regression model, $\pi(\mathbf{W}; \alpha) = \exp\left(\alpha^{\top}\mathbf{W}\right) / \left(1 + \exp\left(\alpha^{\top}\mathbf{W}\right)\right)$. The estimator of α, say $\hat{\alpha}$, can be obtained by solving $\mathbf{U}^{\alpha}(\alpha) = 0$, where

$$\mathbf{U}^{\alpha}(\alpha) = \sum_{i=1}^{n} \frac{R_{0i} - \pi\left(\mathbf{W}_i; \alpha\right)}{\pi\left(\mathbf{W}_i; \alpha\right)\left\{1 - \pi\left(\mathbf{W}_i; \alpha\right)\right\}} \cdot \frac{\partial \pi\left(\mathbf{W}_i; \alpha\right)}{\partial \alpha^{\top}}.$$

Substitute $\pi\left(\mathbf{W}_i; \hat{\alpha}\right)$ for π_i in (6.5) and denote the resulting score function as $\mathbf{U}_{\mathrm{sw}}^{\beta}(\beta, \hat{\alpha})$. The estimator of β, say $\hat{\beta}_{\mathrm{sw}}(\hat{\alpha})$, is the solution to $\mathbf{U}_{\mathrm{sw}}^{\beta}(\beta, \hat{\alpha}) = 0$.

If $\pi(\mathbf{W}; \alpha)$ is correctly specified, then under some regular conditions, $\hat{\beta}_{\mathrm{sw}}(\hat{\alpha})$ is consistent, and its limiting distribution is

$$\sqrt{n}\left(\hat{\beta}_{\mathrm{sw}}(\hat{\alpha}) - \beta\right) \xrightarrow{d} N\left(0, \Sigma^{-1}\left(\Sigma_{\mathrm{sw}} - I_{\alpha\beta} I_{\alpha}^{-1} I_{\alpha\beta}^{\top}\right)\Sigma^{-1}\right), \tag{6.8}$$

where $I_{\alpha\beta} = \mathbb{E}\left\{\pi^{-1}(\mathbf{W}; \alpha)\,\mathbf{M}_{\tilde{Z}}\partial\pi(\mathbf{W}; \alpha)/\partial\alpha^{\top}\right\}$ and $I_{\alpha} = \mathbb{E}\left\{\mathbf{U}^{\alpha}(\alpha)^{\otimes 2}\right\}$, which can be estimated consistently by $\hat{I}_{\alpha\beta} = -\partial\mathbf{U}_{\mathrm{sw}}^{\beta}\left(\hat{\beta}_{\mathrm{sw}}(\hat{\alpha}), \hat{\alpha}\right)/\partial\alpha^{\top}$ and $\hat{I}_{\alpha} = -\partial\mathbf{U}^{\alpha}(\hat{\alpha})/\partial\alpha^{\top}$.

Because $I_{\alpha\beta} I_{\alpha}^{-1} I_{\alpha\beta}^{\top}$ is positive definite, we may obtain a more asymptotically efficient estimator of β by estimating π. Also note that if π depends only on $\mathbf{Z}_1$, then π is predictable, and $I_{\alpha\beta}$ is equal to 0 by the martingale theory. Therefore, we suggest estimating π even when π is known and using all possible predictors, especially the outcome variables, in the parametric model. Explicitly, we may include all complete covariates, the censoring indicator, the significant polynomial of the follow-up time, and the significant interactions between the foregoing variables in the π model. An elaborate model can also help prevent model misspecification.

6.4.1.3 *Model π Nonparametrically* Since it is not guaranteed that an assumed parametric model is correct, Qi et al. (2005) used a nonparametric regression,

Nadaraya–Watson estimation, to model the missing mechanism. That is, when **W** is categorical,

$$\hat{\pi}(\mathbf{w}) = \frac{\sum_{i=1}^{n} R_{0i} I\left(\mathbf{W}_i = \mathbf{w}\right)}{\sum_{i=1}^{n} I\left(\mathbf{W}_i = \mathbf{w}\right)};$$

when **W** contains continuous elements,

$$\hat{\pi}(\mathbf{w}) = \frac{\sum_{i=1}^{n} R_{0i} K_h\left(\mathbf{W}_i = \mathbf{w}\right)}{\sum_{i=1}^{n} K_h\left(\mathbf{W}_i = \mathbf{w}\right)},$$

where K_h is a kernel function with bandwidth h. If **W** contains both the categorical and continuous elements, which is more likely the case in practice, then we can stratify the sample by the categorical elements of **W** and carry out Nadaraya–Watson estimation with the continuous elements of **W** in each strata. Replacing π with $\hat{\pi}$ in (6.5), the resulting estimator is denoted by $\hat{\beta}_{\text{sw}}(\hat{\pi})$. Qi et al. (2005) showed that

$$\sqrt{n}\left(\hat{\beta}_{\text{sw}}(\hat{\pi}) - \beta\right) \xrightarrow{d} N\left(0, \Sigma^{-1}\Sigma_{\text{sw}}^{\hat{\pi}}(\pi)\Sigma^{-1}\right), \tag{6.9}$$

where $\Sigma_{\text{sw}}^{\hat{\pi}}(\pi) = \Sigma + \Sigma_{\text{sw}}^{*}(\pi)$ and $\Sigma_{\text{sw}}^{*}(\pi) = \mathbb{E}\left\{\left(\pi^{-1} - 1\right)\operatorname{var}\left(\mathbf{M}_{\tilde{Z}}|\mathbf{W}\right)\right\}$.

It can be seen that $\Sigma_{\text{sw}}^{\hat{\pi}}(\pi) = \Sigma_{\text{sw}}(\pi) - \mathbb{E}\left\{\left(\pi^{-1} - 1\right)\mathbf{M}_{\tilde{Z}}^{0\otimes 2}\right\}$, where $\mathbf{M}_{\tilde{Z}}^{0} = \mathbb{E}\left(\mathbf{M}_{\tilde{Z}}|\mathbf{W}\right)$. This means that with the consistent estimator $\hat{\pi}$, $\hat{\beta}_{\text{sw}}(\hat{\pi})$ is a consistent estimator for the true parameter β and has a smaller asymptotic variance than $\hat{\beta}_{\text{sw}}(\pi)$. Estimating π nonparametrically enables $\hat{\beta}_{\text{sw}}(\hat{\pi})$ to use the available data more efficiently than $\hat{\beta}_{\text{sw}}(\pi)$, which is based only on the completely observed subjects. This indicates that the efficiency of the estimator $\hat{\beta}_{\text{sw}}(\pi)$ can be improved by estimating π even if π is known. Furthermore, $\operatorname{var}\left(\mathbf{M}_{\tilde{Z}}|\mathbf{W}\right)$ is a nonincreasing function of the set of variables contained in **W**. So, the more variables that **W** contains, that is, the more information that we use to estimate π, the smaller the $\operatorname{var}\left(\mathbf{M}_{\tilde{Z}}|\mathbf{W}\right)$ and the smaller the asymptotic variance of $\hat{\beta}_{\text{sw}}(\hat{\pi})$. Therefore, efficiency gains are possible if π is estimated as a function of not only the variables on which missingness depends, but also the variables that may be correlated with the covariates subject to missingness. Such $\hat{\pi}$ is consistent for the true π, so the resulting $\hat{\beta}_{\text{sw}}(\hat{\pi})$ is consistent for β.

Similarly, the asymptotic variance of $\hat{\beta}_{\text{sw}}(\hat{\pi})$ is estimated by $\hat{\Sigma}^{-1}\hat{\Sigma}_{\text{sw}}^{\hat{\pi}}(\pi)\hat{\Sigma}^{-1}$, where $\hat{\Sigma}$ is obtained from (6.6) by replacing π with $\hat{\pi}$, and $\hat{\Sigma}_{\text{sw}}^{\hat{\pi}}(\pi) = \hat{\Sigma} + \hat{\Sigma}_{\text{sw}}^{*}(\pi)$. Here, $\hat{\Sigma}_{\text{sw}}^{*}(\pi)$ is obtained as follows. Recall that $\mathbf{M}_{\tilde{Z}}^{0} = \mathbb{E}\left\{\mathbf{M}_{\tilde{Z}}|\mathbf{W}\right\}$. Estimating $\mathbf{M}_{\tilde{Z}}$ by replacing π in (6.7) with $\hat{\pi}$ and $\mathbf{M}_{\tilde{Z}}^{0}$ by the Nadaraya–Watson estimator $\hat{\mathbf{M}}_{\tilde{Z},i}^{0}$, we can obtain an estimator for $\Sigma_{\text{sw}}^{*}(\pi)$:

$$\hat{\Sigma}_{\text{sw}}^{*}(\pi) = \frac{1}{n}\sum_{i=1}^{n}\frac{R_{0i}\left(1 - \hat{\pi}_i\right)}{\hat{\pi}_i^2}\left(\hat{\mathbf{M}}_{\tilde{Z},i} - \hat{\mathbf{M}}_{\tilde{Z},i}^{0}\right)^{\otimes 2}.$$

6.4.2 Augmented Weighted Estimation

Under the MAR assumption, Robins et al. (1994) proposed to add an augmentation term to a simple weighted estimating equation in standard regression models for uncensored data. Based on this idea, we could propose the augmented weighted estimation for the Cox regression with missing covariates.

Discussions about the simple weighted estimator in Section 6.4.1 suggest that $\mathbf{W}$ should contain as many variables as possible. So we let $\mathbf{W} = \left\{X, \Delta, \mathbf{Z}_1\right\}$. Then, it can be easily seen that the augmented versions of $\mathbf{s}^{(k)}(\beta, t)$ and $\mathbf{e}(\beta, t)$ are given by

$$\mathbf{S}_{\text{aw}}^{(k)}(\beta, \pi, t) = \frac{1}{n}\sum_{j=1}^{n}\left[\frac{R_{0j}}{\pi_j}Y_j(t)\exp\left(\beta^\top \mathbf{Z}_j\right)\mathbf{Z}_j^{\otimes k} + \left(1 - \frac{R_{0j}}{\pi_j}\right)Y_j(t)\,\mathbb{E}\left\{\exp\left(\beta^\top \mathbf{Z}_j\right)\mathbf{Z}_j^{\otimes k}|X_j, \Delta_j, \mathbf{Z}_{1j}\right\}\right]$$

and

$$\mathbf{E}_{\text{aw}}(\beta, \pi, t) = \mathbf{S}_{\text{aw}}^{(1)}(\beta, \pi, t)\,/\,S_{\text{aw}}^{(0)}(\beta, \pi, t),$$

respectively. Then by replacing $\mathbf{Z}_i$ in (6.1) with its augmented weighted version and $\mathbf{E}$ with $\mathbf{E}_{\text{aw}}$, the augmented weighted score function is obtained as follows:

$$\mathbf{U}_{\text{aw}}(\beta, \pi) = \sum_{i=1}^{n}\Delta_i\left\{\left(\begin{array}{c}\mathbf{Z}_{1i}\\ \frac{R_{0i}}{\pi_i}\mathbf{Z}_{2i} + \left(1 - \frac{R_{0i}}{\pi_i}\right)\mathbb{E}\left[\mathbf{Z}_{2i}|X_i, \Delta_i, \mathbf{Z}_{1i}\right]\end{array}\right) - \mathbf{E}_{\text{aw}}\left(\beta, \pi, X_i\right)\right\}. \tag{6.10}$$

If π, $\mathbb{E}\left[\mathbf{Z}_2|X, \Delta, \mathbf{Z}_1\right]$, and $\mathbb{E}\left\{\exp\left(\beta^\top \mathbf{Z}\right)\mathbf{Z}^{\otimes k}|X, \Delta, \mathbf{Z}_1\right\}$ for a given β are known, then the solution to $\mathbf{U}_{\text{aw}}(\beta, \pi) = 0$, denoted by $\hat{\beta}_{\text{aw}}(\pi, \mathbb{E})$, is a consistent estimator of β. Qi et al. (2005) called $\mathbf{U}_{\text{aw}}(\beta, \pi)$ the fully augmented weighted score function because $\mathbf{e}(\beta, t)$ is estimated by the augmented weighted method, and the whole score function is the corresponding augmented weighted version of $\sum_{i=1}^{n}\pi_i^{-1}R_{0i}\Delta_i\left\{\mathbf{Z}_i - \mathbf{E}_{\text{aw}}\left(\beta, \pi, X_i\right)\right\}$. Under some regular conditions, it is shown that

$$n^{1/2}\left(\hat{\beta}_{\text{aw}}(\pi, \mathbb{E}) - \beta\right)\stackrel{d}{\longrightarrow} N\left(0, \Sigma^{-1}\Sigma_{\text{aw}}(\pi)\Sigma^{-1}\right),$$

where $\Sigma_{\text{aw}}(\pi) = \Sigma + \Sigma_{\text{aw}}^{*}(\pi)$ and $\Sigma_{\text{aw}}^{*}(\pi) = E\left\{\left(\pi^{-1} - 1\right)\text{var}\left(\mathbf{M}_{\tilde{Z}}|X, \Delta, \mathbf{Z}_1\right)\right\}$. With true π, the augmented estimator is more efficient than the simple weighted estimator. However, when π is estimated conditional on $\{X, \Delta, \mathbf{Z}_1\}$, they are equally efficient.

It is more likely the case in practice that the selection probability π and/or the conditional expectations $\mathbb{E}\left(\cdot|X,\Delta,\mathbf{Z}_1\right)$ are unknown. As discussed in Section 6.4.1, π can be estimated parametrically or nonparametrically. For the estimation of the conditional expectations, it can be implemented more easily by the nonparametric Nadaraya–Watson estimation. If π and/or the conditional expectations $\mathbb{E}\{\cdot|X,\Delta,\mathbf{Z}_1\}$ are estimated nonparametrically, and the resulting estimates are substituted into the augmented weighted score function, then the corresponding estimators of β are denoted by $\hat{\beta}_{\text{aw}}(\hat{\pi},\mathbb{E})$, $\hat{\beta}_{\text{aw}}\left(\pi,\hat{\mathbb{E}}\right)$, and $\hat{\beta}_{\text{aw}}\left(\hat{\pi},\hat{\mathbb{E}}\right)$. Qi et al. (2005) showed that $\hat{\beta}_{\text{aw}}(\pi,\mathbb{E})$, $\hat{\beta}_{\text{aw}}(\hat{\pi},\mathbb{E})$, $\hat{\beta}_{\text{aw}}\left(\pi,\hat{\mathbb{E}}\right)$, and $\hat{\beta}_{\text{aw}}\left(\hat{\pi},\hat{\mathbb{E}}\right)$ share the same asymptotic properties; that is, they are all asymptotically normal at rate root n with mean 0 and variance matrix $\Sigma^{-1}\Sigma_{\text{aw}}(\pi)\Sigma^{-1}$. This indicates that nonparametric estimation of the selection probabilities and the conditional expectations do not affect the asymptotic properties of the augmented estimators.

A consistent estimator of the variance of $\hat{\beta}_{\text{aw}}(\pi,\mathbb{E})$ can be obtained similarly to the simple weighted estimator. Let

$$\mathbf{V}_{\text{aw}}(\beta,\pi,t)=\frac{\mathbf{S}^{(2)}_{\text{aw}}(\beta,\pi,t)}{S^{(0)}_{\text{aw}}(\beta,\pi,t)}-\mathbf{E}_{\text{aw}}(\beta,\pi,t)^{\otimes 2},$$

$$\mathrm{d}\hat{\Lambda}_0^{\text{aw}}(t,\hat{\pi})=\frac{1}{n}\frac{\sum_{j=1}^{n}\mathrm{d}N_j(t)}{S^{(0)}_{\text{aw}}\left(\hat{\beta}_{\text{aw}}\left(\hat{\pi},\hat{\mathbb{E}}\right),\hat{\pi},t\right)},$$

$$\mathrm{d}\hat{M}_i(t)=dN_i(t)-Y_i(t)\exp\left\{\hat{\beta}^{\top}_{\text{aw}}\left(\hat{\pi},\hat{\mathbb{E}}\right)\mathbf{Z}_i\right\}\mathrm{d}\hat{\Lambda}_0^{\text{aw}}(t,\hat{\pi}),$$

and

$$\hat{\mathbf{M}}_{\tilde{Z},i}=\int_0^{\tau}\left[\mathbf{Z}_i-\mathbf{E}^{(1)}_{\text{aw}}\left(\hat{\beta}_{\text{aw}}\left(\hat{\pi},\hat{\mathbb{E}}\right),\hat{\pi},t\right)\right]\mathrm{d}\hat{M}_i(t).$$

Then, Σ and $\Sigma^{*}_{\text{aw}}(\pi)$ are estimated by

$$\hat{\Sigma}=\sum_{i=1}^{n}\Delta_i\mathbf{V}_{\text{aw}}\left(\hat{\beta}_{\text{aw}}\left(\hat{\pi},\hat{\mathbb{E}}\right),\hat{\pi},X_i\right)$$

and

$$\hat{\Sigma}^{*}_{\text{aw}}(\pi)=\frac{1}{n}\sum_{i=1}^{n}\frac{R_{0i}\left(1-\hat{\pi}_i\right)}{\hat{\pi}_i^2}\left(\hat{\mathbf{M}}_{\tilde{Z},i}-\hat{\mathbf{M}}^{0}_{\tilde{Z},i}\right)^{\otimes 2},$$

respectively, where $\hat{\mathbf{M}}^{0}_{\tilde{Z},i}$ is obtained using the Nadaraya–Watson estimator. We can obtain that the estimator of $\hat{\beta}_{\text{aw}}(\pi,\mathbb{E})$ is $\hat{\Sigma}^{-1}+\hat{\Sigma}^{-1}\hat{\Sigma}^{*}_{\text{aw}}(\pi)\hat{\Sigma}^{-1}$.

The nonparametric kernel estimation method for the selection probability and the conditional expectations yields estimators that, in the sense of a large sample, are as effective as those with known selection probability and conditional expectations. However, when the dimension of $\mathbf{Z}_1$ is large, the kernel function is required to have high order and large bandwidth. This might make the resulting estimator perform badly in a finite sample. So, in this case, one might prefer to model the selection probability and the conditional expectations parametrically. This is done by Wang and Chen (2001) and Xu et al. (2009). They suggested augmented weighted estimating equations, which yield consistent estimators if either the selection probability or the conditional expectation is modeled correctly. Wang and Chen (2001) suggested an expectation–maximization (EM) algorithm to obtain estimates from their proposed estimating equations. The EM algorithm may not be feasible with a complex conditional distribution of missing covariates given the observed data. So, in the following, we only introduce the method proposed by Xu et al. (2009).

As we did in Section 6.4.1.2, when the selection probability is modeled parametrically, we denote $\hat{\pi}_i = \pi(\mathbf{W}_i; \hat{\alpha})$ with $\mathbf{W}_i$ containing X_i, Δ_i, and $\mathbf{Z}_{1i}$. When the conditional expectations are unknown, for simplicity, we assume that the distribution of censoring time does not depend on unobserved covariates, so that only the cumulative baseline hazard $\Lambda_0(t)$ and the conditional distribution of $\mathbf{Z}_2$ given $\mathbf{Z}_1$ are needed to construct the conditional distribution of the missing covariates given the observed data. We estimate the unknown $\Lambda_0(t)$ by

$$\hat{\Lambda}_0(t) = \frac{1}{n}\sum_{i=1}^{n}\int_0^t \frac{R_{0i}(s)/\pi\left(\mathbf{W}_i;\alpha\right)}{S_{\text{sr}}^{(0)}(\alpha,\beta,s)}\mathrm{d}N_i(s).$$

Let $S\left(\mathbf{Z}_2|\mathbf{Z}_1;\gamma\right)$ be the score function from the conditional distribution $P\left(\mathbf{Z}_2|\mathbf{Z}_1;\gamma\right)$. Then, the estimator of γ, say $\hat{\gamma}$, is obtained by solving $\mathbf{U}_{\text{aw}}^{\gamma}(\hat{\alpha},\gamma) = 0$, where

$$\begin{aligned}\mathbf{U}_{\text{aw}}^{\gamma}(\hat{\alpha},\gamma) = \frac{1}{n}\sum_{i=1}^{n}\Bigg\{&\frac{R_{0i}}{\pi\left(\mathbf{W}_i;\hat{\alpha}\right)}S\left(\mathbf{Z}_{2i}|\mathbf{Z}_{1i};\gamma\right)\\ &+\left(1-\frac{R_{0i}}{\pi\left(\mathbf{W}_i;\hat{\alpha}\right)}\right)\hat{\mathbb{E}}\left[S\left(\mathbf{Z}_{2i}|\mathbf{Z}_{1i};\gamma\right)\right]\Bigg\}.\end{aligned}$$

Based on $\hat{\alpha}$, $\hat{\gamma}$, and $\hat{\Lambda}_0(t)$, we could obtain the estimates of $\mathbb{E}(\cdot|\mathbf{W};\beta,\gamma)$, denoted by $\hat{\mathbb{E}}(\cdot|\mathbf{W};\beta,\hat{\gamma})$, and then the estimated $\mathbf{S}_{\text{aw}}^{(k)}$ is denoted by $\hat{\mathbf{S}}_{\text{aw}}^{(k)}$. Replacing $\mathbf{S}_{\text{aw}}^{(k)}$ and $\mathbb{E}\left(\mathbf{Z}_{2i}|X_i,\Delta_i,\mathbf{Z}_{1i}\right)$ in (6.10) with $\hat{\mathbf{S}}_{\text{aw}}^{(k)}$ and $\hat{\mathbb{E}}\left(\mathbf{Z}_{2i}|X_i,\Delta_i,\mathbf{Z}_{1i};\beta,\hat{\gamma}\right)$, respectively, the resulting score function, say $\mathbf{U}_{\text{aw}}^{\beta}(\beta,\hat{\alpha},\hat{\gamma})$, yields an estimator of β. Xu et al. (2009) showed that the estimation of the nuisance parameters α, γ, and $\Lambda_0(t)$ does not affect the asymptotic results of the estimator of β. So, all the augmented estimators have the same asymptotic distribution as $\hat{\beta}_{\text{aw}}(\pi,\mathbb{E})$.

6.4.3 Reweighting Estimators

The inverse probability weighted estimators are consistent under the MAR assumption. However, when the selection probability is small for some observations, heavy weights will be assigned to those observations and the variance of the IPW estimators may be inflated. In addition, even small bias for the small selection probability would cause large bias for the score function, and thus the estimator of β. To mitigate this problem, the reweighting method could be used. Inspired by the concept of risk set sampling (Langholz and Goldstein, 1996), Xu et al. (2009) proposed the simple reweighting and augmented reweighting estimators. The basic idea is that at the first step each complete subject is weighted by the inverse of the observation probability to correct the selection bias due to missingness, and then a simpler selection probability is imposed on each subject in the pseudo-unbiased sample generated from the first step. Xu et al. (2009) suggested using the marginal observation probability, given a risk set at time t, as the imposed simpler selection probability. Thus, each subject in the same risk set is assigned an identical selection probability $\pi^*(t)$. The weight for each complete observation in the same risk set becomes $\pi^*(t)/\pi_i$. To avoid an extra auxiliary model, we use the empirical estimator to estimate $\pi^*(t)$:

$$\hat{\pi}^*(t) = \frac{\sum_{i=1}^n R_{0i} Y_i(t)}{\sum_{i=1}^n Y_i(t)}.$$

In the following, we only introduce the simple reweighting and augmented reweighting estimators by assuming that the selection probability and the conditional expectations given X, Δ, and $\mathbf{Z}_1$ are known. Analogous results are obtained when the selection probability and conditional expectations are estimated.

The simple reweighting versions of $\mathbf{s}^k(\beta, \pi, t)$ and $\mathbf{e}(\beta, \pi, t)$ are

$$\mathbf{S}_{\text{sr}}^{(k)}(\beta, \pi, t) = n^{-1} \sum_{j=1}^{n} R_{0j} \frac{\hat{\pi}^*(t)}{\pi_j} Y_j(t) \exp\left(\beta^\top \mathbf{Z}_j\right) \mathbf{Z}_j^{\otimes k}$$

and

$$\mathbf{E}_{\text{sr}}(\beta, \pi, t) = \mathbf{S}_{\text{sr}}^{(1)}(\beta, \pi, t) / S_{\text{sr}}^{(0)}(\beta, \pi, t),$$

respectively. Because $\hat{\pi}^*(t)$ does not depend on j, after factoring out the $\hat{\pi}^*(t)$ in $\mathbf{S}_{\text{sr}}^{(1)}$ and $S_{\text{sr}}^{(0)}$, we have

$$\mathbf{E}_{\text{sr}}(\beta, \pi, t) = \mathbf{S}_{\text{sw}}^{(1)}(\beta, \pi, t) / S_{\text{sw}}^{(0)}(\beta, \pi, t) = \mathbf{E}_{\text{sw}}(\beta, \pi, t).$$

Therefore, the simple reweighting score function is

$$\mathbf{U}_{\text{sr}}(\beta, \pi) = \sum_{i=1}^{n} R_{0i} \frac{\hat{\pi}^*(X_i)}{\pi_i} \Delta_i \left\{\mathbf{Z}_i - \mathbf{E}_{\text{sw}}\left(\beta, \pi, X_i\right)\right\}. \tag{6.11}$$

The corresponding estimator of β is denoted by $\hat{\beta}_{\text{sr}}(\pi)$.

Denote

$$\mathbf{M}^*_{\tilde{Z}} = \int_0^\tau \pi^*(t)\,[\mathbf{Z} - \mathbf{e}(\beta, t)]\,\mathrm{d}M(t),$$

$$\Sigma^* = \int_0^\tau \pi^*(t)\,\mathbf{v}(\beta, t)\,s^{(0)}(\beta, t)\,\mathrm{d}\Lambda_0(t),$$

and $\Sigma_{\mathrm{sr}}(\pi) = \mathbb{E}\left\{\pi^{-1}\mathbf{M}^{*\otimes 2}_{\tilde{Z}}\right\}$. Xu et al. (2009) showed that, under some regular conditions, $\hat{\beta}_{\mathrm{sr}}(\pi)$ is consistent and the asymptotic distribution of $n^{1/2}\hat{\beta}_{\mathrm{sr}}(\pi)$ is

$$\sqrt{n}\left(\hat{\beta}_{\mathrm{sr}}(\pi) - \beta\right) \xrightarrow{d} N\left(0, (\Sigma^*)^{-1}\Sigma_{\mathrm{sr}}(\pi)(\Sigma^*)^{-1}\right).$$

The consistent estimator of the asymptotic variance can be obtained similarly to the simple weighted estimator. When π is estimated by the parametric and nonparametric methods, the results are similar: just remember to replace Σ with Σ^* and $\mathbf{M}_{\tilde{Z}}$ with $\mathbf{M}^*_{\tilde{Z}}$ in (6.8) and (6.9), respectively.

For the augmented reweighting estimation, we define

$$\begin{aligned}\mathbf{S}^{(k)}_{\mathrm{ar}}(\beta, \pi, t) = \frac{1}{n}\sum_{j=1}^{n}\Bigg[& R_{0j}\frac{\hat{\pi}^*(t)}{\pi_j}Y_j(t)\exp\left(\beta^\top \mathbf{Z}_j\right)\mathbf{Z}_j^{\otimes k} \\ &+ \left(\pi_j - R_{0j}\right)\frac{\hat{\pi}^*(t)}{\pi_j}Y_j(t)\,\mathbb{E}\left\{\exp\left(\beta^\top \mathbf{Z}_j\right)\mathbf{Z}_j^{\otimes k} | X_j, \Delta_j, \mathbf{Z}_{1j}\right\}\Bigg],\end{aligned}$$

for $k = 0$, 1, and 2. We can see that

$$\mathbf{E}_{\mathrm{ar}}(\beta, \pi, t) = \mathbf{S}^{(1)}_{\mathrm{ar}}(\beta, \pi, t) \,/\, S^{(0)}_{\mathrm{ar}}(\beta, \pi, t) = \mathbf{E}_{\mathrm{aw}}(\beta, \pi, t).$$

Therefore, the augmented reweighting score function is

$$\begin{aligned}&\mathbf{U}_{\mathrm{ar}}(\beta, \pi) \\ &= \sum_{i=1}^{n}\Delta_i\hat{\pi}^*(X_i)\left\{\begin{pmatrix}\mathbf{Z}_{1i} \\ \frac{R_{0i}}{\pi_i}\mathbf{Z}_{2i} + \left(1 - \frac{R_{0i}}{\pi_i}\right)\mathbb{E}\left[\mathbf{Z}_{2i} | X_i, \Delta_i, \mathbf{Z}_{1i}\right]\end{pmatrix} - \mathbf{E}_{\mathrm{aw}}\left(\beta, \pi, X_i\right)\right\}.\end{aligned}$$

The corresponding estimator of β is denoted by $\hat{\beta}_{\mathrm{ar}}$. It is shown by Xu et al. (2009) that

$$\sqrt{n}\left(\hat{\beta}_{\mathrm{ar}} - \beta\right) \xrightarrow{d} N\left(0, \left(\Sigma^*\right)^{-1}\left\{\Sigma_{\mathrm{sr}}(\pi) - \mathbb{E}\left[\pi^{-1}(1-\pi)\mathbf{M}^{*0\otimes 2}_{\tilde{Z}}\right]\right\}\left(\Sigma^*\right)^{-1}\right),$$

where $\mathbf{M}^{*0}_{\tilde{Z}} = \mathbb{E}\left(\mathbf{M}^{*0}_{\tilde{Z}}|X, \Delta, \mathbf{Z}_1\right)$. Similar to the augmented weighted estimators, whether or not the selection probability and the conditional expectations are estimated, the resulting estimators have the same asymptotic properties.

6.5 IMPUTATION METHODS

In the presence of missing data, it is natural to impute the missing data. One type of imputation is the conditional expectation imputation, which is introduced in Section 6.5.1. The other type of imputation is to impute the missing data by some values, and consider the resulting imputed data set as complete data, to which statistical methods for complete data can be applied. This imputation method is introduced in Section 6.2.

6.5.1 Imputation by Conditional Expectations

Paik and Tsai (1997) proposed to impute the conditional expectation of the statistic involving missing covariates, given the available information. Recall that if the data are completely observed, the solution of $\mathbf{U}_{\mathrm{f}}(\beta) = 0$ is a consistent estimator of β, where $\mathbf{U}_{\mathrm{f}}(\beta)$ is as defined in (6.1). This is because the summand of $\mathbf{U}_{\mathrm{f}}(\beta)$ satisfies $\mathbb{E}\left\{\Delta_i\left[\mathbf{Z}_i - \mathbf{E}(\beta, X_i)\right]\right\} = 0$. So, when the missing covariates are present, we could impute $\mathbb{E}\left(\mathbf{Z}_{2i}|X_i, \Delta_i, \mathbf{Z}_{1i}\right)$ for the missing $\mathbf{Z}_{2i}$ and $\mathbb{E}\left\{\mathbf{E}(\beta, X_i)\,|X_i, \Delta_i, \mathbf{Z}_{1i}\right\}$ for $\mathbf{E}(\beta, X_i)$. This will also produce a consistent estimator of β because

$$\mathbb{E}\left\{\Delta_i\left[\mathbb{E}\left(\mathbf{Z}_{2i}|X_i, \Delta_i, \mathbf{Z}_{1i}\right) - \mathbb{E}\left\{\mathbf{E}(\beta, X_i)\,|X_i, \Delta_i, \mathbf{Z}_{1i}\right\}\right]\right\} = 0.$$

Due to the special structure of $\mathbf{E}(\beta, X_i)$ in the Cox model, we have

$$\mathbb{E}\left\{\mathbf{E}(\beta, X_i)\,|X_i, \Delta_i, \mathbf{Z}_{1i}\right\} \approx \frac{\mathbb{E}\left\{\mathbf{S}^{(1)}\left(\beta, X_i\right)|X_i, \Delta_i, \mathbf{Z}_{1i}\right\}}{\mathbb{E}\left\{S^{(0)}\left(\beta, X_i\right)|X_i, \Delta_i, \mathbf{Z}_{1i}\right\}}.$$

So, what we have to do is to impute $\mathbb{E}\left(\mathbf{Z}_{2i}|X_i, \Delta_i, \mathbf{Z}_{1i}\right)$ for $\mathbf{Z}_{2i}$ if $\mathbf{Z}_{2i}$ is missing, and $\mathbb{E}\left\{\mathbf{S}^{(1)}\left(\beta, X_i\right)|X_i, \Delta_i, \mathbf{Z}_{1i}\right\}$ and $\mathbb{E}\left\{S^{(0)}\left(\beta, X_i\right)|X_i, \Delta_i, \mathbf{Z}_{1i}\right\}$ for $\mathbf{S}^{(1)}\left(\beta, X_i\right)$ and $S^{(0)}\left(\beta, X_i\right)$, respectively. Furthermore, since some of the elements of $\mathbf{Z}_{2i}$ may be missing and the others are observed, we only impute the conditional expectation for the missing elements. Paik and Tsai (1997) considered that $\mathbf{Z}_{1i}$ is categorical. Then, if the lth element of $\mathbf{Z}_{2i}$, say Z_{2li}, is missing, they propose to estimate $\mathbb{E}\left(Z_{2li}|X_i, \Delta_i, \mathbf{Z}_{1i}\right)$ by

$$\hat{\mathbb{E}}\left(Z_{2li}|X_i, \Delta_i, \mathbf{Z}_{1i}\right) = \frac{\sum_{k=0}^{K}\sum_{j=1}^{n} I\left\{X_j \in J_k, X_i \in J_k, \mathbf{Z}_{1j} = \mathbf{Z}_{1i}\right\}\Delta_j R_{(p+l)j} Z_{2lj}}{\sum_{k=0}^{K}\sum_{j=1}^{n} I\left\{X_j \in J_k, X_i \in J_k, \mathbf{Z}_{1j} = \mathbf{Z}_{1i}\right\}\Delta_j R_{(p+l)j}},$$

where

$$J_k = \left(c_k, c_{k+1}\right], \quad 0 = c_0 < c_1 < c_2 < \cdots < c_K < c_{K+1} = \tau.$$

The other conditional expectations are estimated similarly. Then, after imputing the statistics involving missing values, we have a score function, which produces a consistent estimator of β under some regular conditions. The variance of the estimator can be obtained by bootstrap.

6.5.2 Multiple Imputation

To account for the uncertainty due to the missing data, several data sets are generated by imputing the missing data for several times. This is the so-called multiple imputation. Statistical inference is performed on each imputed data set, and the results are pooled together by Rubin's rule. It is typical that multiple imputation is more efficient than the complete-case analysis for regression analysis when covariates have missing values. Since it is more understandable than the other missing data methods for practical users and is implemented by STATA, SAS, and R, multiple imputation is widely used in practice. We first describe multiple imputation in Section 6.5.2.1 for the case in which there is only one covariate having missing values; that is, Z_2 is a variable. If there are more than one covariate having missing values, then multiple imputation by chained equations (MICE) can be applied, which is introduced in Section 6.5.2.2.

6.5.2.1 Multiple Imputation for One Covariate For univariate Z_2, multiple imputation is performed using a regression model for Z_2 on other variables, including $\mathbf{Z}_1$, X, and Δ. Suppose the imputation model $p\left(Z_2|X, \Delta, \mathbf{Z}_1; \alpha\right)$ is parameterized by α. Formally, multiple imputation involves drawing values of the missing data Z_2 from the predictive distribution:

$$p\left(Z_2|X, \Delta, \mathbf{Z}_1\right) = \int p\left(Z_2|X, \Delta, \mathbf{Z}_1; \alpha\right) p\left(\alpha|X, \Delta, \mathbf{Z}_1\right) \mathrm{d}\alpha,$$

where $p\left(\alpha|X, \Delta, \mathbf{Z}_1\right)$ is the Bayesian posterior distribution of α. Normally, we should specify a prior distribution for α. However in practice, with implicit vague priors, the values of Z_2 may be obtained by (i) fitting the model $p\left(Z_2|X, \Delta, \mathbf{Z}_1; \alpha\right)$ to the cases with observed Z_2, yielding an estimate $\hat{\alpha}$ with estimated covariance matrix S_α; (ii) drawing a value of α, α^*, from its posterior, perhaps approximated as $N\left(\hat{\alpha}, S_\alpha\right)$; and (iii) drawing values of Z_2 from $p\left(Z_2|X, \Delta, \mathbf{Z}_1; \alpha^*\right)$. This substantially simplifies the algorithm and reduces the computation complexity.

It is important to include the censoring time X and event indicator Δ in the imputation model. Commonly, X is included in the imputation model with $\log X$, but the rationale is not clear. White and Royston (2009) showed and recommended that, instead of the logarithm of survival time, imputation should be based on the cumulative hazard to the survival time, which if not known, can be replaced by its

Nelson–Aalen estimator. They found that for small covariate effects and/or small cumulative incidence, a suitable model for imputing binary or normal Z_2 is a logistic or linear regression on the event indicator Δ, the cumulative baseline hazard $\Lambda_0(T)$, and the other covariates $\mathbf{Z}_1$. In the case of a single binary covariate Z_2, the imputing model for Z_2 is exactly a logistic regression on Δ and $\Lambda_0(T)$. The result is shown below.

First, with binary Z_2 and no $\mathbf{Z}_1$, we have

$$\text{logit}\, p\left(Z_2 = 1|X, \Delta\right) = \alpha_0 + \alpha_1 \Delta + \alpha_2 \Lambda_0(X),$$

where $\Lambda_0(X)$ is the cumulative baseline hazard $\int_0^X \lambda_0(t)\,\mathrm{d}t$. Second, with binary Z_2 and categorical $\mathbf{Z}_1$, we can obtain that for the most general exposure model $\text{logit}\, p\left(Z_2 = 1|\mathbf{Z}_1\right) = \zeta_{\mathbf{Z}_1}$,

$$\text{logit}\, p\left(Z_2 = 1|X, \Delta, \mathbf{Z}_1\right) = \alpha_0 + \alpha_1 \Delta + \alpha_2 \Lambda_0(X) + \alpha_{3\mathbf{Z}_1} + \alpha_{4\mathbf{Z}_1} \Lambda_0(X),$$

where $\alpha_{3\mathbf{Z}_1}$ and $\alpha_{4\mathbf{Z}_1}$ represent a set of dummy variables with their coefficients. In other cases, we can only obtain approximate results. For binary Z_2 with more general $\mathbf{Z}_1$, we use a Taylor series approximation for $\exp\left(\beta_1^\top \mathbf{Z}_1\right)$ that is valid when $\beta_1^\top \mathbf{Z}_1$ has a small variance. By assuming the exposure model $\text{logit}\, p\left(Z_2 = 1|\mathbf{Z}_1\right) = \zeta_0 + \zeta_1^\top \mathbf{Z}_1$, we have

$$\text{logit}\, p\left(Z_2 = 1|X, \Delta, \mathbf{Z}_1\right) \approx \alpha_0 + \alpha_1 \Delta + \alpha_2 \Lambda_0(X) + \alpha_3^\top \mathbf{Z}_1.$$

Adding an interaction term $\alpha_4^\top \mathbf{Z}_1 \Lambda_0(X)$ can improve the accuracy of the approximation. Furthermore, if the analyst believed that a particular transformation of $\mathbf{Z}_1$ was needed for predicting Z_2, then this transformation should be entered in the imputation model.

For normal Z_2, we use a Taylor series approximation for $\exp\left(\beta_1^\top \mathbf{Z}_1 + \beta_2 Z_2\right)$ that is valid when $\beta_1^\top \mathbf{Z}_1 + \beta_2 Z_2$ has a small variance. Suppose the exposure model is $Z_2|\mathbf{Z}_1 \sim N\left(\zeta_0 + \zeta_1^\top \mathbf{Z}_1, \sigma^2\right)$. Then by using a first-order approximation, we have

$$Z_2|X, \Delta, \mathbf{Z}_1 \approx N\left(\alpha_0 + \alpha_1 \Delta + \alpha_2 \Lambda_0(X) + \alpha_3^\top \mathbf{Z}_1, \sigma^2\right),$$

and again adding an interaction term $\alpha_4^\top \mathbf{Z}_1 \Lambda_0(X)$ improves the accuracy of the approximation. Departures from the above model will be most significant when both $\text{var}\left(\beta_2 Z_2\right)$ and $\Lambda_0(x) \exp\left(\beta_1^\top \bar{\mathbf{Z}}_1 + \beta_2 \bar{Z}_2\right)$ are large.

6.5.2.2 Multiple Imputation for Multivariate MICE starts by filling in missing values arbitrarily and then applies the univariate method for each incomplete variable in turn, using the current imputed values of $\mathbf{Z}$ when drawing new values of $\mathbf{Z}_2$, and vice versa. The procedure is iterated until convergence, which often requires fewer than 10 cycles. An alternative noniterative procedure is available if the data are monotonically missing. In the data example, we will use mice package of R.

6.6 NONPARAMETRIC MAXIMUM LIKELIHOOD ESTIMATION

Chen and Little (1999) proposed the NPML estimation. Although the NPML method requires heavier computation workload, it can be applied in situations where the weighted estimating equation methods cannot be used due to the lack of complete cases in some survival time intervals.

Recall that $\mathbf{R}$ is the missingness indicator diagonal matrix, that is, its jth element of the diagonal is equal to 1 if the jth covariate of a subject is observed, and 0 otherwise. We further define $\mathbf{R}$ as a function in $\mathbb{R}^r$; that is, for $\mathbf{z} \in \mathbb{R}^r$, $\mathbf{R}(\mathbf{z})$ is a vector formed by the jth element of $\mathbf{z}$ where the jth element of diagonal of $\mathbf{R}$ is equal to 1. So, $\mathbf{R}_i(\mathbf{Z}_i)$ is the value of the observed covariates of the ith subject. The observed data are $(X_i, \Delta_i, \mathbf{R}_i(\mathbf{Z}_i), \mathbf{R}_i)$, $i = 1, \ldots, n$.

To write the likelihood for the observed data, we assume that the censoring distribution does not depend on the unobserved covariates, that is, the censoring distribution satisfies

$$G(C, \mathbf{Z}) = G(C, \mathbf{R}(\mathbf{Z})).$$

We also assume that the missingness mechanism is

$$P(\mathbf{R}|X, \Delta, \mathbf{Z}) = P(\mathbf{R}|X, \Delta, \mathbf{R}(\mathbf{Z})), \tag{6.12}$$

which is more general than the usual MAR assumption (6.4). Suppose that $\mathbf{Z}$ has a density $f(\mathbf{z}, \theta)$ with respect to a known measure. Then, the likelihood for the observed data $(X_i, \Delta_i, \mathbf{R}_i(\mathbf{Z}_i))$, $i = 1, \ldots, n$, is

$$\prod_{i=1}^{n} \lambda^{\Delta_i}(X_i) \int_{\mathbf{R}_i(\mathbf{z})=\mathbf{R}_i(\mathbf{Z}_i)} \exp\left(\Delta_i \beta^\top \mathbf{z}\right) \exp\left\{-\Lambda(X_i) \exp\left(\beta^\top \mathbf{z}\right)\right\} \lambda_g^{\Delta_i}(X_i, \mathbf{z})$$
$$\times \left(1 - G(X_i, \mathbf{z})\right) f(\mathbf{z}, \theta)\, d\mathbf{z} \prod_{i=1}^{n} P\left(\mathbf{R}_i | X_i, \Delta_i, \mathbf{R}_i(\mathbf{Z}_i)\right),$$

where λ is the baseline hazard density for the survival times and λ_g is the hazard density for the censoring times. If the distributions of censoring time and the missing data indicator vector do not depend on the parameters of interest, namely, the regression parameters and the baseline hazard, then the likelihood up to a constant factor is

$$\begin{aligned}
&L(\Lambda, \beta, \theta \mid X, \Delta, \mathbf{R}(\mathbf{Z})) \\
&= \prod_{i=1}^{n} \lambda^{\Delta_i}(X_i) \int_{\mathbf{R}_i(\mathbf{z})=\mathbf{R}_i(\mathbf{Z}_i)} \exp\left(\Delta_i \beta^{\top} \mathbf{z}\right) \exp\left\{-\Lambda(X_i) \exp\left(\beta^{\top} \mathbf{z}\right)\right\} f(\mathbf{z}, \theta)\, \mathrm{d}\mathbf{z}.
\end{aligned} \tag{6.13}$$

It is worth noting that the maximum of this likelihood over the parameter space

$$\left\{(\lambda, \beta, \theta) : \theta \in \mathbb{R}^k, \beta \in \mathbb{R}^r, \lambda(t) \geq 0 \text{ on } (0, +\infty)\right\}$$

does not exist. We require that the cumulative baseline hazard is a step function with jumps at the observed follow-up times. Then, (6.13) can be maximized over the parameter space:

$$\left\{(\Lambda, \beta, \theta) : \theta \in \mathbb{R}^k, \beta \in \mathbb{R}^r, \Lambda(t) \text{ is an increasing step function on } (0, +\infty)\right\}.$$

The resulting estimator is called a nonparametric maximum likelihood estimator. The EM-type algorithms are employed to solve the maximization problem.

6.7 TRANSFORMATION MODEL

Note that the Cox model is a special case of transformation models. Here, we give a short introduction on the application of the nonparametric maximum-likelihood method to the transformation models. Chen and Little (2001) considered the model

$$\log(\Lambda(T)) = -\beta^{\top}\mathbf{Z} + \varepsilon,$$

where $\Lambda(\cdot)$ is an unspecified monotonic transformation function. Actually, the method can deal with the model

$$P(T > t \mid \mathbf{Z}) = \phi(\mathbf{Z}, \theta, \Lambda(t)),$$

where θ contains the parameter β and the parameter related to the distribution of ε and ϕ is a known function.

Suppose that the covariates $\mathbf{Z}$ follow a parametric distribution denoted by $H_{\gamma}(\mathbf{z})$, where γ is a q-dimensional parameter. Then the likelihood for the observed data is

$$\prod_{i=1}^{n} \int_{\mathbf{R}_i(\mathbf{z})=\mathbf{R}_i(\mathbf{Z}_i)} p\left(\mathbf{R}_i \mid X_i, \Delta_i, \mathbf{Z} = \mathbf{z}\right) p\left(X_i, \Delta_i \mid \mathbf{Z} = \mathbf{z}, \theta, \Lambda\right) \mathrm{d}H_{\gamma}(\mathbf{z}).$$

Under the missing mechanism (6.12), the parts of likelihood describing the missing data mechanism and the censoring mechanism can be dropped from the full likelihood. The remaining part can be written as

$$\prod_{i=1}^{n}\int_{\mathbf{R}_i(\mathbf{z})=\mathbf{R}_i(\mathbf{Z}_i)}\left[\varphi\left\{\mathbf{z},\theta,\Lambda\left(X_i\right)\right\}\right]^{\Delta_i}\left[\phi\left\{\mathbf{z},\theta,\Lambda\left(X_i\right)\right\}\right]^{1-\Delta_i}\mathrm{d}H_\gamma\left(\mathbf{z}\right)\left\{\mathrm{d}\Lambda\left(X_i\right)\right\}^{\Delta_i}, \tag{6.14}$$

where φ is the derivative of $-\phi$ with respect to the argument Λ. For the proportional hazards regression model, maximizing the likelihood (6.14) with respect to (θ, Λ) was proposed by Chen and Little (1999). To reduce the computational complexity, here a different approach is employed. The basic idea is to first transform the parameters $(\theta, \gamma, \Lambda)$ into (θ, γ, S), where S denotes the marginal survival function without covariate. Then, we factor the likelihood for the new set of parameters into two parts: the first part is the conditional likelihood of $\mathbf{R}(\mathbf{Z})$ given (X, Δ), which contains all the information about (θ, γ) and a small amount of information about S, and the second part is the marginal likelihood of (X, Δ), which contains most information about S. So, we ignore the part of information about S contained in the first part of likelihood and maximize the second part of likelihood to estimate S. The resulting estimate of S is substituted into the first part of likelihood, which is maximized to obtain estimate of (θ, γ).

Specifically, note that

$$S\left(t\right)=\int\phi\left\{\mathbf{z},\theta,\Lambda\left(t\right)\right\}\mathrm{d}H_\gamma\left(\mathbf{z}\right),$$

and under weak conditions, the inverse transformation

$$\Lambda\left(t\right)=\upsilon\left(\theta,\gamma,S\left(t\right)\right)$$

exists and we have

$$\frac{\mathrm{d}S}{\mathrm{d}\Lambda}\left(t\right)=-\int\varphi\left\{\mathbf{z},\theta,\Lambda\left(t\right)\right\}\mathrm{d}H_\gamma\left(\mathbf{z}\right).$$

Then, the likelihood (6.14) can be factored as

$$L\left(\theta,\gamma,S\right)=L_1\left(\theta,\gamma,S\right)L_2\left(S\right),$$

where $L_1(\theta, \gamma, S) = \prod_{i=1}^{n} p\left(\mathbf{R}_i\left(\mathbf{Z}_i\right) | X_i, \Delta_i, \theta, \gamma, S\right)$ with

$$p\left(\mathbf{R}_i\left(\mathbf{Z}_i\right) | X_i, \Delta_i, \theta, \gamma, S\right) = \left(\frac{\int_{\mathbf{R}_i(\mathbf{z})=\mathbf{R}_i(\mathbf{Z}_i)} \varphi\left(\mathbf{z}, \theta, v\left\{\theta, \gamma, S\left(X_i\right)\right\}\right) \mathrm{d}H_\gamma(\mathbf{z})}{\int \varphi\left(\mathbf{z}, \theta, v\left\{\theta, \gamma, S\left(X_i\right)\right\}\right) \mathrm{d}H_\gamma(\mathbf{z})}\right)^{\Delta_i}$$
$$\times\left(\frac{\int_{\mathbf{R}_i(\mathbf{z})=\mathbf{R}_i(\mathbf{Z}_i)} \phi\left(\mathbf{z}, \theta, v\left\{\theta, \gamma, S\left(X_i\right)\right\}\right) \mathrm{d}H_\gamma(\mathbf{z})}{S\left(X_i\right)}\right)^{1-\Delta_i}$$

and

$$L_2(S) = \prod_{i=1}^{n} S^{1-\Delta_i}\left(X_i\right)\left\{\mathrm{d}S\left(X_i\right)\right\}^{\Delta_i}.$$

Note that the first part of likelihood, L_1, contains little information about S. We estimate S by maximizing the second part of likelihood L_2 with respect to S. Typically the resulting estimate of S is the Kaplan–Meier estimator $\hat{S}$ based on data (X_i, Δ_i). Then, we estimate (θ, γ) by maximizing $L_1(\theta, \gamma, \hat{S})$. The latter procedure is called the profile conditional likelihood method. Maximizing L_1 with respect to (θ, γ) is a routine calculation. In particular, the EM algorithm can be applied to simplify the computations. The resulting estimates of (θ, γ), denoted by $(\hat{\theta}, \hat{\gamma})$, are asymptotically normal.

6.8 DATA EXAMPLE: PATHWAYS STUDY

Now we turn to the Pathways data. As we have seen, there are missing covariates for some subjects, so we use the complete-case analysis, enhanced complete-case analysis, simple weighted method, augmented weighted method, simple reweighting estimation, augmented reweighting estimation, and multiple imputation with chained equations to deal with the missing covariates problem under the Cox regression model. Table 6.1 reports the estimates of the regression coefficients β and their standard errors. The p-values for testing whether the coefficients are zero are also shown in the table.

It can be seen that the complete-case analysis and the enhanced complete-case analysis have somewhat different results. Age has a significant effect on death in both methods, but the impact differs considerably: the coefficient of age for the complete-case analysis is 0.057, whereas that for the enhanced complete-case analysis is 0.094. Similar results are obtained for the covariate smoking. The covariates major depression, eGFR, and microalbuminuria are significant in both methods, and the complete-case estimates of their coefficients are close to the enhanced complete-case estimates. The complete-case analysis shows that the education and mean duration of diabetes are significant, but the enhanced complete-case analysis reveals that they are not. The HbA1c is not significant in both methods.

TABLE 6.1 The Regression Coefficients and Their Standard Errors and *p*-Values for the Cox Regression Model

	Complete-Case			ECC		
	Coefficient	SE	*p*-Value	Coefficient	SE	*p*-Value
Age	0.057	0.009	0.000	0.094	0.009	0.000
Education	−0.472	0.147	0.001	−0.205	0.144	0.154
Smoking	0.558	0.265	0.035	0.718	0.285	0.012
Mean duration of diabetes	0.018	0.007	0.007	0.008	0.006	0.245
HbA1c	0.038	0.053	0.465	0.011	0.055	0.834
Major depression	0.434	0.202	0.032	0.472	0.213	0.027
eGFR	−0.015	0.004	0.000	−0.018	0.004	0.000
Microalbuminuria	0.672	0.148	0.000	0.565	0.158	0.000
	SW			**SR**		
	Coefficient	SE	*p*-Value	Coefficient	SE	*p*-Value
Age	0.063	0.008	0.000	0.064	0.010	0.000
Education	−0.361	0.136	0.008	−0.359	0.162	0.027
Smoking	0.587	0.260	0.024	0.581	0.312	0.062
Mean duration of diabetes	0.020	0.006	0.001	0.020	0.008	0.008
HbA1c	0.042	0.051	0.403	0.042	0.061	0.493
Major depression	0.458	0.197	0.020	0.459	0.236	0.052
eGFR	−0.013	0.005	0.003	−0.013	0.005	0.013
Microalbuminuria	0.744	0.145	0.000	0.745	0.171	0.000
	AW			**AR**		
	Coefficient	SE	*p*-Value	Coefficient	SE	*p*-Value
Age	0.088	0.006	0.000	0.088	0.008	0.000
Education	−0.143	0.103	0.167	−0.143	0.124	0.248
Smoking	0.832	0.199	0.000	0.821	0.240	0.001
Mean duration of diabetes	0.017	0.005	0.001	0.017	0.006	0.004
HbA1c	0.055	0.037	0.138	0.054	0.044	0.227
Major depression	0.345	0.149	0.021	0.347	0.179	0.052
eGFR	0.004	0.004	0.237	0.004	0.004	0.328
Microalbuminuria	0.632	0.112	0.000	0.639	0.132	0.000
	MICE					
	Coefficient	SE	*p*-Value			
Age	0.065	0.006	0.000			
Education	−0.202	0.107	0.057			
Smoking	0.534	0.201	0.008			
Mean duration of diabetes	0.014	0.005	0.006			
HbA1c	0.017	0.039	0.672			
Major depression	0.280	0.149	0.060			
eGFR	−0.015	0.003	0.000			
Microalbuminuria	0.594	0.132	0.000			

As we have indicated, the complete-case analysis and the enhanced complete-case analysis rely on the MCAR assumption. A simple logistic regression shows that there are some significant factors for the missingness. This means that the MCAR is not satisfied. In this case, the results of the complete-case analysis and the enhanced complete-case analysis might not be reliable. Hence we also carried out the analysis by the weighted methods and multiple imputation. For the weighted methods, the selection probability and the conditional expectations are all estimated by the nonparametric Nadaraya–Watson estimation. It can be seen from Table 6.1 that the simple weighted and augmented weighted estimates of β are close to the simple reweighting and augmented reweighting estimates, respectively. However the reweighting methods have larger standard errors. Therefore we might have more trust in the results of the nonreweighting methods.

The age, smoking status, mean duration of diabetes, major depression, and microalbuminuria are significant in both the simple weighted and augmented weighted methods. However, education and eGFR are significant in the simple weighted method, whereas they are not in the augmented weighted method. In the multiple imputation with chained equations, education is nearly significant, and the eGFR is very significant as shown in the simple weighted method.

6.9 CONCLUDING REMARKS

In this chapter, we introduce some methods for dealing with the statistical inference for the survival data with the missing covariates. The complete-case analysis and the enhanced complete-case analysis produce consistent estimators only under the MCAR missing mechanism. In our data example, they produce different results. As indicated by Lin and Ying (1993), the enhanced complete-case analysis is more efficient than the complete-case analysis. So, we might prefer to use the enhanced complete-case analysis instead of the complete-case analysis. However, we should first check whether the MCAR assumption is satisfied. If not, we should use the other methods.

Theoretical investigations show that the inverse probability weighted methods (simple weighted and simple reweighting methods) have the same efficiency as the augmented inverse probability weighted methods (augmented weighted and augmented reweighting methods) if the selection probability is estimated. We also note that the simple weighted method only requires to model the selection probability, whereas the augmented weighted method needs to model several conditional expectations in addition to the selection probability and may introduce more estimation errors in the finite sample. Our application to the Pathways data confirms this conclusion. So, the inverse probability weighted methods might be preferred than the augmented inverse probability weighted methods.

The reweighting methods are introduced to mitigate the overweighting problem for the nonreweighting methods. But it cannot always produce more efficient estimators than the nonweighting methods. So, in practice, we may report the results obtained from both the reweighting and nonweighting methods and check which one is more reliable.

Multiple imputation with chained equation is widely used in practice. However, whether it is a proper imputation and whether it produces consistent estimation have not been examined in theoretical investigations. We could report its results together with other methods, and by comparing the results, we can see which one is reasonable.

CHAPTER 7

NONIGNORABLE MISSINGNESS

7.1 INTRODUCTION

In the previous chapters, the focus was on data that were missing at random (MAR); that is, the missing mechanism is explained by the observed variables only. Under the MAR assumption, the observed-data likelihood can be written as a product of the missing mechanism part and the outcome model part. So the maximum likelihood estimation of the outcome model could be done separately from estimating the missing mechanism; this property is called "ignorability."

However, the MAR assumption cannot be verified using the observed data alone without assuming a parametric model. It is very possible that the missing mechanism depends on the missing data itself, or on some other unobserved variables that are associated with the missing data. For example, in a cancer clinical trial, suppose the outcome variable is a biomarker reflecting the treatment effect, and is only observed at the end of the study. Subjects may decide to drop out before the endpoint because the treatment seems to be ineffective, which leads to a violation of the MAR assumption. In the case of missing not at random (MNAR), the observed-data likelihood involves the joint distribution of the outcome and the missing indicator, which cannot be separated from each other.

Similar to the MAR assumption, MNAR cannot be tested from the observed data alone nonparametrically. The decision to consider MNAR models should be mainly based on scientific knowledge about the reasons for missing data, rather than the

Applied Missing Data Analysis in Health Sciences, First Edition. Xiao-Hua Zhou, Chuan Zhou, Danping Liu, and Xiaobo Ding.

observed data set alone. On the other hand, if one is uncertain about whether the MAR assumption holds, it is recommended that MNAR models always be fitted as a sensitivity analysis. Under a parametric framework for the MNAR model, the validity of the MAR assumption could be tested. Although this is a rather *ad hoc* test, the results could inform the data analyst about the strength of possible ''nonignorability''.

Early in the data collection stage, efforts should be made to reduce the possibility of MNAR missingness. First, it is crucial for the researchers to investigate the reasons for missing data carefully and try to collect all possible covariates that predict the missingness. Particularly in longitudinal studies, collecting time-varying covariate information often facilitates modeling the missingness. Second, it is helpful to define the study population carefully and to focus on a more homogeneous target population. For example, if the outcome of interest is measured at 1-year follow-up, it is more reasonable to exclude the subjects who are very ill and may not survive for one year. Third, in the cross-sectional studies, one should avoid a long time lag between the exposure and outcome variable. If such a time lag is critical in answering the scientific questions (e.g., examining a long-term treatment effect), a longitudinal design should be used instead.

In this chapter, we focus on four types of methods to deal with MNAR data: likelihood-based methods, Bayesian methods, multiple imputation methods, and estimating equation methods. Little (1995) identified two classes of likelihood-based nonignorable models: selection models and pattern mixture models. The former assume a missing indicator distribution given the outcome variable, whereas the latter assume different outcome models stratified by the missing patterns. The shared random effects model (Wu and Carroll, 1988; Follmann and Wu, 1995), in a broader sense, belongs to the selection model class. It is used only when the data are longitudinal: the missing mechanism is not directly conditional on the outcome, but on some random effects, and the same set of random effects also affects the longitudinal outcome process. Therefore, the dependence between the outcome and the missing indicator is induced by latent variables. We also discuss other likelihood approaches that do not directly belong to Little's classification, for example, the mixed effects hybrid model proposed by Yuan and Little (2009). The likelihood-based methods have the efficiency advantage when the distributional assumptions are valid. However, for more complicated missing data patterns, evaluating the observed-data likelihood may be too cumbersome. In Section 7.2, we discuss the selection model for cross-sectional data. In Section 7.3, we introduce the models for longitudinal data, but limit our scope only to the missing outcome, which could be either monotone or nonmonotone missing.

When the missing data pattern becomes more complicated, nonlikelihood-based methods are often preferred. Huang et al. (2005) developed a full Bayesian method for estimating regression models with MNAR covariates, and examined the properties of the priors used for the missing data model under various conditions. Multiple imputation (MI), which is a frequentist version of the Bayesian data augmentation method, was proposed by Glynn et al. (1993) to deal with nonignorable nonresponse under both selection and pattern mixture models. Another general approach for nonignorable dropouts is based on estimating equations (Rotnitzky et al., 1998). The key idea is to construct a selection model and then weight the estimating equations using the selection probabilities. A sensitivity analysis framework was suggested by Rotnitzky et al. (2001). We introduce these nonlikelihood-based methods in Sections 7.4–7.6.

7.2 CROSS-SECTIONAL DATA: SELECTION MODEL

7.2.1 Missing Outcome

Let Y_i be the outcome for subject i and X_i be the vector of covariates, for $i = 1, \ldots, N$. Let R_i be the missing indicator of Y_i. Suppose interest lies in the regression model of Y_i on X_i, given by a generalized linear model:

$$E(Y_i | X_i) = g(X_i'\beta), \tag{7.1}$$

where g is the link function and β is the vector of regression coefficients. Let $f(Y_i | X_i)$ denote the density of Y_i given X_i, which belongs to the exponential family.

With a nonignorable missing outcome, the likelihood involves the joint distribution of Y_i and R_i. Under a selection model framework, the joint distribution is factored out as

$$P(Y_i, R_i | X_i) = P(R_i | Y_i, X_i) f(Y_i | X_i).$$

The first part $P(R_i | Y_i, X_i)$ can be modeled by a binary regression model,

$$P(R_i | Y_i, X_i) = h(X_i'\gamma + \alpha Y_i), \tag{7.2}$$

where h is a link function and α is called the ''nonignorable parameter.'' When $\alpha = 0$, the model reduces to ignorable missingness; a larger absolute value of α indicates stronger nonignorability.

The log-likelihood can then be written as

$$\begin{aligned}\log L(\alpha, \beta, \gamma) = \sum_{i=1}^{N} [R_i\{\log P(R_i = 1 | Y_i, X_i) + \log f(Y_i | X_i)\} \\ +(1 - R_i) \log \int P(R_i = 0 | Y_i, X_i) f(Y_i | X_i) \mathrm{d}Y_i].\end{aligned}$$

The integration is with respect to the missing Y_i when $R_i = 0$. If Y_i is categorical, the integral simplifies to the summation over all possible values of Y_i; if Y_i is continuous, we can use the Gaussian–Hermite quadrature to evaluate the integral. In general, the parameter β is not separable from α and γ in the likelihood.

7.2.2 Missing Covariate

For simplicity, we consider the situation when only one covariate is missing. We change the notation slightly from the previous section. Let $X_i = (X_{1i}, X'_{2i})'$ denote the covariates, where X_{1i} is subject to missingness and X_{2i} is complete. Let R_i be the missing indicator of X_{1i}. The same regression model (7.1) is of primary interest. For estimation, we need to write the joint distribution of Y_i, X_{1i}, and R_i given X_{2i}:

$$P(Y_i, X_{1i}, R_i | X_{2i}) = P(R_i | Y_i, X_i) f(Y_i | X_i) f(X_{1i} | X_{2i}).$$

An additional model assumption of $X_{1i} | X_{2i}$ needs to be made, so that we can integrate out the missing X_{1i} from the observed-data likelihood. Assume that $X_{1i} | X_{2i}$ can also be specified using a generalized linear model (GLM):

$$E(X_{1i} | X_{2i}) = g(X'_{2i}\eta).$$

The selection model and the response model are still given by (7.1) and (7.2), respectively.

The log-likelihood is

$$\begin{aligned}
&\log L(\alpha, \beta, \gamma, \eta) \\
&= \sum_{i=1}^{N} R_i \{\log P(R_i = 1 | Y_i, X_i) + \log f(Y_i | X_i) + \log f(X_{1i} | X_{2i}) \\
&\quad + (1 - R_i) \log \int P(R_i = 0 | Y_i, X_i) f(Y_i | X_i) f(X_{1i} | X_{2i}) \mathrm{d}X_{1i}\}.
\end{aligned}$$

The integral with respect to X_{1i} needs to be evaluated numerically for continuous X_{1i}, and reduces to a summation for categorical X_{1i}.

When the number of missing covariates increases, the likelihood-based approach quickly becomes too complicated because of the multivariate integration. In this case, we recommend the use of multiple imputation, which is introduced in Section 7.5.

7.3 LONGITUDINAL DATA WITH DROPOUT

7.3.1 Notation

Assume that N subjects are collected in the sample. Each subject is to be observed at m fixed time points, denoted by $t_1, \ldots, t_m$. Let $Y_i^{\mathrm{c}} = (Y_{i1}, \ldots, Y_{im})'$ denote the complete vector of responses for subject i. Let R_i denote the dropout time, and n_i denote the number of observation times before dropout time R_i. The observed outcome vector is $Y_i = (Y_{i1}, \ldots, Y_{in_i})'$. Let $X_i = (X_{i1}, \ldots, X_{im})'$ denote the corresponding covariate matrix for subject i. Note that X_i may often include observation time t_{ij}.

7.3.2 Overview

When data are not missing at random, the likelihood involves the joint distribution of the response and the missingness mechanism, $f(Y, R| X)$. Three types of methods can be used, depending on how the joint distribution is factored: selection models, pattern mixture models, and shared random effects models.

The selection models factor out the joint distribution by a full data model and the missing mechanism model:

$$f(Y, R| X) = f(Y| X)f(R| Y, X).$$

The advantage of this framework is that the regression model of interest, $f(Y| X)$, is directly modeled and estimated. The pattern mixture models assume a full data model stratified by the missingness patterns:

$$f(Y, R| X) = f(Y| R, X)f(R| X).$$

This model treats the full data as a mixture over dropout times, which makes the inference on $f(Y| X)$ less straightforward. The shared random effects model assumes that a common random effect term, b, is associated with the outcome and the missingness mechanism:

$$f(Y, R| X) = \int f(Y| X, b)f(R| X, b)\mathrm{d}F(b| X).$$

The dependence of both the response process and the missingness process is solely captured by the shared random effects. The model often requires more intensive computation to evaluate the integral in the likelihood function.

7.3.3 Pattern Mixture Model

7.3.3.1 Identification Constraints The pattern mixture models are based on the factorization

$$f(Y, R| X) = f(Y| R, X)f(R| X).$$

The measurement model $f(Y| R, X)$ describes the dependence of the outcome Y on the missing indicator R, that is, the "strength" of nonignorable missingness. A saturated model specifies a different joint distribution of Y for each missingness pattern, but not all the parameters can be identified without imposing further constraints. For $R_i = t$, the full data density is given by

$$f_t(Y_{i1}, \ldots, Y_{im}) = f_t(Y_{i1}, \ldots, Y_{it})f_t(Y_{i,t+1}, \ldots, Y_{im}| Y_{i1}, \ldots, Y_{it}). \tag{7.3}$$

The first term is clearly identifiable because $Y_{i1}, \ldots, Y_{it}$ are all observed, but the second term is not identifiable.

Jansen and Molenberghs (2010) discussed the general strategy of specifying the constraints for the pattern mixture model. The idea is to relate the second term in (7.3) back to the patterns with the observed data for $Y_{i,t+1}, \ldots, Y_{im}$. A general expression is given by

$$f_t(Y_{is} | Y_{i1}, \ldots, Y_{i,s-1}) = \sum_{j=s}^{m} w_{sj} f_j (Y_{is} | Y_{i1}, \ldots, Y_{i,s-1}), \tag{7.4}$$

where w_{sj} is a weight summing up to 1 ($\sum_{j=s}^{m} w_{sj} = 1$). For the jth pattern ($R_i = j$) with $j \geq s$, the outcome Y_{is} is observed, so their densities lend information to the unobserved density in the sth pattern ($R_i = s$).

There are several special cases in the literature that use the setup of (7.4). Little (1993) assumed that the conditional density beyond the dropout time borrows information only from the completers; that is,

$$f_t(Y_{is} | Y_{i1}, \ldots, Y_{i,s-1}) = f_m(Y_{is} | Y_{i1}, \ldots, Y_{i,s-1}),$$

for $s = t + 1, \ldots, m$. This is called the complete-case missing value (CCMV) restriction. This method is useful when the majority of subjects are completers, and only a small proportion of subjects have various dropout patterns. Curran et al. (2004) proposed a neighboring case missing value (NCMV) restriction, where the nearest identifiable pattern is used:

$$f_t(Y_{is} | Y_{i1}, \ldots, Y_{i,s-1}) = f_s(Y_{is} | Y_{i1}, \ldots, Y_{i,s-1}).$$

Another special case is the available case missing value (ACMV), as proposed by Molenberghs et al. (1998). They derived the weight as

$$w_{sj} = \frac{\pi_j f_j (Y_{i1}, \ldots, Y_{i,s-1})}{\sum_{l=s}^{m} \pi_l f_l(Y_{i1}, \ldots, Y_{i,s-1})},$$

where π_j is the fraction of observations in pattern j. It is also shown that the ACMV restriction is equivalent to the MAR assumption.

7.3.3.2 Generalized Linear Pattern Mixture Model The key assumption in the pattern mixture model is the distribution $f(Y | R, X)$. Several authors focused on the multivariate normal distribution (Little and Wang, 1996; Molenberghs et al., 1998; Hogan et al., 2004). Hogan and Laird (1997) and Fitzmaurice and Laird (2000) considered a more general case of any type of outcome and proposed a generalized linear model for estimating $f(Y | R, X)$. We introduce their methods in this section.

Consider the form of a generalized linear model:

$$E(Y_{ij} | R_i, X_{ij}) = g(W_{ij}' \beta), \tag{7.5}$$

where g is a known link function, W_{ij} is the design vector, and β is the parameter vector. The design vector W_{ij} is expanded by X_{ij} and R_i, and may also incorporate the interaction terms. It is important to note that β may not always be identifiable, unless some parametric constraint is specified. To illustrate this, we consider an example of $m = 3$, and X_{ij} only includes the observation time t_j. The dropout indicator R_i can take the value of t_2, t_3, or t_4, indicating those individuals who made the first observation only, first two observations only, or all three observations, respectively. A saturated model includes the full interaction terms between t_j and R_i:

$$E(Y_{ij}|R_i = t_k, X_{ij}) = g(\beta_0^k + \beta_1^k t_j),$$

for $k = 2, 3, 4$. This model is not identifiable because for subjects with $R_i = t_2$, only one observation at t_1 is available; thus, the slope β_1^2 could not be estimated. We have to make additional assumptions on the structure of $E(Y_{ij}|R_i, X_{ij})$ so that subjects who dropped out can borrow information from the completers. For example, by assuming that the slope of time is constant for different dropout patterns, the model could be identified:

$$E(Y_{ij}|R_i = t_k, X_{ij}) = g(\beta_0^k + \beta_1 t_j).$$

Or, we can assume that the slope of time depends linearly on the dropout time:

$$E(Y_{ij}|R_i, X_{ij}) = g\{\beta_0 + \beta_1 t_j + \beta_2 R_i + \beta_3(t_j \times R_i)\},$$

which is identifiable as well. The pattern mixture model (7.5) can be estimated by generalized estimating equations (Fitzmaurice and Laird, 2000). Hogan and Laird (1997) proposed a generalized linear mixed effects model as an alternative way to estimate $f(Y|R, X)$. These models are readily implemented in standard software packages, as reviewed in Chapter 5.

The next step is to estimate the dropout probability, $P(R_i|X_i)$. Since there are no missing data involved in this model, one can simply employ a continuation ratio model, similar to that in Section 5.6.1, but without including Y_i in the model. If there are only categorical covariates or no covariates at all, we can use the sample proportion of dropout to estimate the dropout probability.

One last issue to note is that the target of inference is not β, but the conditional mean $E(Y_{ij}|X_{ij})$, which needs to be averaged over the dropout patterns:

$$E(Y_{ij}|X_{ij}) = \sum_{k=2}^{m+1} P(R_i = t_k|X_{ij})g(W_{ij}'\beta).$$

Eventually, $E(Y_{ij}|X_{ij})$ may not follow the generalized linear form of X_{ij}, so it is often difficult to quantify the effect size of X_{ij}. However, if X_{ij} is categorical (e.g., in randomized clinical trials, X_{ij} may only include the treatment variable), we can compute the marginal mean of the outcome for each covariate level and make the comparisons.

7.3.3.3 *Latent Dropout Class Model* In the situation where there is a relatively large number of follow-up times, some dropout patterns may have relatively few subjects. It could be challenging to estimate the parameters in those strata, unless strong restrictions are made on the parameters. To address this problem, Roy (2003) and Roy and Daniels (2008) proposed a latent dropout class model. They assume that the dropout patterns are clustered into C ($C < m$) latent classes; the outcome model is not specified for each actual dropout time, but for each latent dropout class.

Let O_i be the latent dropout class to which subject i belongs. Assume that O_i can take values from 1 to C, with larger values indicating later dropout time. The latent class probability is then modeled using an ordinal regression model:

$$P(O_i \leq k | R_i) = h(R_i; \lambda_k),$$

for $k = 1, \ldots, C - 1$, where h is a monotone link function and λ_k is the parameter vector. For example, we can specify

$$h(R_i; \lambda_k) = \frac{\exp(\lambda_{0k} + \lambda_1 R_i)}{1 + \exp(\lambda_{0k} + \lambda_1 R_i)},$$

which assumes that the slope of R_i is constant across the classes and the intercept is different. The probability of being in a dropout class can be expressed as

$$P(O_i = k | R_i) = \begin{cases} h(R_i; \lambda_1), & \text{for } k = 1, \\ h(R_i; \lambda_k) - h(R_i; \lambda_{k-1}), & \text{for } k > 1. \end{cases}$$

This model is easily extended to the situation when additional covariates X_i also affect O_i.

The key assumption in the latent dropout class model is that within the same latent class, the dropout becomes missing at random. In other words, the dependence of Y_{ij} on R_i is only through O_i:

$$E(Y_{ij} | R_i, O_i, X_{ij}) = E(Y_{ij} | O_i, X_{ij}).$$

Unlike Section 7.3.3.2 where the mean response model (7.5) is enough, here we need to specify the multivariate joint distribution $Y_i | R_i, O_i, X_i$. Consider the generalized linear mixed effects model:

$$E(Y_{ij} | O_i, X_i, b_i) = g(W'_{ij}\beta + Z'_{ij} b_i),$$

where W_{ij} is the design vector expanded by O_i, X_i, and possibly the interaction terms, b_i is the vector of random effects, and Z_{ij} is the design vector of random effects. Write $f(Y_{ij} | O_i, X_i, b_i)$ to denote the conditional density of Y_{ij}. The pattern mixture likelihood can be calculated by integrating over the random effects and the latent class:

$$L_i(Y_i|R_i, X_i) = \sum_{k=1}^{K} \int f(Y_{ij}|O_i = k, X_i, b_i)P(O_i = k|R_i)\mathrm{d}F(b_i).$$

As discussed in Chapter 5, the integral in the likelihood is often intractable unless Y_{ij} and b_i both follow normal distribution. In practice, we need to employ numerical methods to evaluate the integration, such as the Gaussian–Hermite quadrature. The parameter estimation can be obtained by maximizing the log-likelihood $\sum_{i=1}^{N} L_i(Y_i|R_i, X_i)$. The dropout probability $P(R_i|X_i)$ can be estimated using the methods described in the previous section. The target of inference, $E(Y_{ij}|X_{ij})$, can be written as

$$E(Y_{ij}|X_{ij}) = \sum_{O_i}\sum_{R_i} P(O_i|R_i, X_i)P(R_i|X_i) \int E(Y_{ij}|O_i, X_i, b_i)\mathrm{d}F(b_i).$$

7.3.4 Selection Model

For longitudinal data, the selection model factors out the joint distribution $f(Y, R|X)$ as $f(Y|X)f(R|Y, X)$. Diggle and Kenward (1994) gave the general selection model framework to deal with nonignorable dropout data. We introduce their approach in this section.

Let $H_{id} = \{Y_{i1}, \ldots, Y_{i,d-1}, X_i\}$ denote the sequence of outcome measurements and covariates for subject i before time d. We could then model the probability of dropping out at time t_d, for example, using a logistic regression:

$$\text{logit}\left\{P(R_i = d|H_{id}, Y_{id})\right\} = \beta_0 + \sum_{j=1}^{d} \beta_j Y_{ij} + \beta'_{d+1}X_i. \tag{7.6}$$

Here, the dropout probability at time t_d relies on the outcome variable Y_{id} at the same time. Therefore, the dropout process is nonignorable if $\beta_d \neq 0$.

The next step is to specify the outcome model. Diggle and Kenward (1994) only considered the continuous outcome, and assumed a multivariate normal distribution:

$$Y_i \sim N(X_i\theta, V_i). \tag{7.7}$$

It is further assumed that $V_i = \upsilon^2 J_i + \sigma^2 G_i + r^2 I_i$, where J_i is an $n_i \times n_i$ matrix of 1's, $G_i = (g_{i,jk})$ is an $n_i \times n_i$ autoregressive matrix with $g_{i,jk} = \exp(-\alpha(t_k - t_k))$, and I_i is the $n_i \times n_i$ identity matrix. Note that $\upsilon^2 J_i$ represents the exchangeable correlation, G_i the serial correlation, and I_i the measurement error. Let us denote $\phi = (\upsilon^2, \sigma^2, r^2, \alpha)'$ as the vector of nuisance parameters for the variance component; denote $f^*_{n_i}(Y_i)$ to be the joint distribution of the observed outcome Y_i given X_i. Then, for a completer, the contribution to the likelihood is

$$f_m^*(Y_i)\prod_{k=2}^{m}\left\{1 - P(R_i = k|H_{ik}, Y_{ik})\right\}. \tag{7.8}$$

For a subject who dropped out at time t_d, the contribution to the likelihood is

$$f_{d-1}^*(Y_i)\left[\prod_{k=2}^{d-1}\left\{1 - P(R_i = k|H_{ik}, Y_{ik})\right\}\right] P(R_i = d|H_{id}). \tag{7.9}$$

The last term in (7.9) does not condition on the unobserved Y_{id}, so it has to be derived from model assumptions (7.6) and (7.7):

$$\begin{aligned} &P(R_i = d|H_{id}) \\ &= \int P(R_i = d|H_{id}, Y_{id}) f^*(Y_{id}|H_{id}) \mathrm{d}Y_{id} \\ &= \int P(R_i = d|H_{id}, Y_{id}) \frac{f^*(Y_{i1}, \ldots, Y_{id}|X_i)}{f^*(Y_{i1}, \ldots, Y_{i,d-1}|X_i)} \mathrm{d}Y_{id}. \end{aligned} \tag{7.10}$$

An implicit assumption here is that the full data $Y_i^{\mathrm{c}} = (Y_{i1}, \ldots, Y_{im})'$ still follow the same multivariate normal distribution as in (7.7). Hence, the conditional distribution $f^*(Y_{id}|\, H_{id})$ is simply a univariate normal density.

The log-likelihood function can be written as

$$\log L(\beta, \theta, \phi) = l_1(\theta, \phi) + l_2(\beta) + l_3(\beta, \theta, \phi),$$

where

$$l_1(\theta, \phi) = \sum_{i=1}^{N} f_{n_i}^*(Y_i),$$

$$l_2(\beta) = \sum_{i=1}^{N}\sum_{k=2}^{n_i} \log\left\{1 - P(R_i = k|H_{ik}, Y_{ik})\right\},$$

and

$$l_3(\beta, \theta, \phi) = \sum_{i:n_i<m} \log P(R_i = n_i + 1|H_{i,n_i+1}).$$

Note that the parameters β, θ, and ϕ cannot be separated in the likelihood, so they have to be estimated jointly. The integral in $l_3(\beta, \theta, \phi)$ needs to be evaluated numerically. Under the special case of ignorable dropout, the third term simplifies to

$$l_3(\beta, \theta, \phi) = \sum_{i:n_i<m} \log \left\{ \frac{\exp(\beta_0 + \sum_{j=1}^{d} \beta_j Y_{ij} + \beta'_{d+1} X_i)}{1 + \exp(\beta_0 + \sum_{j=1}^{d} \beta_j Y_{ij} + \beta'_{d+1} X_i)} \right\},$$

which only involves β. In this case, the parameters of interest θ can be found by maximizing $l_1(\theta, \phi)$.

In principle, it is possible to extend the above model to non-normal outcomes by using a generalized linear mixed model to replace the assumption (7.7). However, it is often difficult to evaluate the integration (7.10) because the joint distribution $f^*(Y_{i1}, \ldots, Y_{id} \mid X_i)$ itself involves integration with respect to the random effects.

7.3.5 Shared Random Effects Model

The shared random effects model was first proposed by Wu and Carroll (1988) to handle the normal longitudinal data with informative censoring. This approach was further extended by Follmann and Wu (1995), which allows for the discrete outcome and more flexible patterns of missing data (both monotone and intermittent missing). We summarize the general model framework in Follmann and Wu (1995).

Suppose that the outcome process is described by a generalized linear mixed model:

$$E(Y_{ij}|X_i, b_i) = g_1(X'_{ij}\beta + Z'_{ij}b_i), \tag{7.11}$$

where g_1 is the link function, X_{ij} and Z_{ij} are, respectively, the design vector of the fixed and random effects, and b_i is the random effects following a multivariate normal distribution $F(\cdot)$. The Y_{ij} terms are conditionally independent given b_i, and their distribution belongs to the exponential family. For normal Y_{ij} as a special case, the conditional independence assumption can be reduced to a multivariate normal assumption of $Y_i|b_i$, which allows for serial correlation within an individual, for example, in (7.7).

The missing data behavior is summarized in the vector $M_i = (M_{i1}, \ldots, M_{iL})'$ for the ith subject. For example, Wu and Carroll (1988) looked at dropout as in the previous sections, so that $L = 1$ and M_{i1} is the actual dropout time R_i; Mori et al. (1994), Dunson et al. (2003), and Chen et al. (2011) considered the informative cluster size, so that $L = 1$ and $M_{i1} = n_i$; others such as Shen and Gao (2007), Tsonaka et al. (2009), and Lin et al. (2010) considered intermittent missing, so that $L = m$ with M_{ij} being the missing indicator for Y_{ij}.

The shared random effects model assumes that the same vector of random effects b_i affects the outcome process and the missing data process at the same time. Therefore, another generalized linear model for M_{il} can be specified as

$$E(M_{il}|X_i, Y_i, b_i) = g_2(W_{il}'\theta_1 + b_i'\theta_2),$$

where g_2 is the link function and W_{il}' is the vector of covariates associated with the missing mechanism, which could include some components of X_i and observed Y_i (such as outcome at the baseline), as well as their interactions. The coefficient θ_2 measures the strength of association between the outcome process and the missing data process, that is, the level of nonignorability. Some components of θ_2 could be set to 0, since not all of the random effects necessarily affect the missingness.

Given the random effects b_i, the outcome Y_i is assumed independent of the missing indicator M_i. Therefore, the observed-data likelihood is obtained as

$$\sum_{i=1}^{N} \log\left\{\int f(Y_i|X_i, b_i) f(M_i|X_i, b_i) \mathrm{d}F(b_i)\right\},$$

which could be evaluated by using the Gaussian–Hermite quadrature. Maximizing the likelihood yields the estimation of parameters β, θ_1, and θ_2.

Follmann and Wu (1995) also discussed the relationship between the shared random effects model and the pattern mixture model. The former can often be approximated by the latter under mild assumptions. For simplicity, the covariates X_i are ignored. If the shared random effects model is correct, we note that

$$\begin{aligned} f(Y_i|M_i) &= \int f(Y_i|M_i, b_i) \mathrm{d}F(b_i|M_i) \\ &= \int f(Y_i|b_i) \mathrm{d}F(b_i|M_i). \end{aligned}$$

The second equality is obtained by using $f(Y_i|M_i, b_i) = f(Y_i|b_i)$, which follows from the independence assumption of Y_i and M_i given b_i. If $F(b_i|M_i)$ can be approximated by a normal distribution not affected by M_i, then $f(Y_i|M_i)$ has the exact form of a generalized linear mixed model and is a pattern mixture model.

In general, $F(b_i|M_i)$ may have a nonzero mean, depending on M_i. As an example, we can express the model $Y_i|M_i$ by modifying (7.11):

$$E(Y_{ij}|M_i, X_i, b_i^*) = g_1[X_{ij}'\beta + Z_{ij}'\left\{b_i^* + E(b_i|M_i)\right\}], \tag{7.12}$$

where b_i^* follows a mean-zero distribution F^* conditional on M_i. Assume that the variance of b_i^* is not affected by M_i, and note that $b_i^* = b_i - E(b_i|M_i)$. We can then specify a linear mean structure for $E(b_i|M_i)$:

$$E(b_i|M_i) = \varpi M_i, \tag{7.13}$$

where ϖ is a parameter matrix with each row being the coefficients for one element of b_i. Then, the conditional model (7.12) becomes

$$\begin{aligned} E(Y_{ij}|M_i, X_i, b_i^*) &= g_1\{X_{ij}'\beta + Z_{ij}'\left(b_i^* + \varpi M_i\right)\} \\ &= g_1(W_{ij}'\beta^* + Z_{ij}'b_i^*), \end{aligned} \tag{7.14}$$

where $W_{ij}' = (X_{ij}', M_i'Z_{ij1}, \ldots, M_i'Z_{ijK})'$, $\beta^* = (\beta', \varpi^{(1)}, \ldots, \varpi^{(K)})'$, and $\varpi^{(k)}$ is the kth row of ϖ. The pattern mixture model (7.14) can be estimated easily using standard software packages. A nice property of this model is that the coefficient β has exactly the same interpretation as that in the regression model of interest (7.11). One can examine the model fit of the conditional random effects distribution $F(b_i|M_i)$ by checking whether the empirical Bayes estimator $\hat{b}_i^*$ is homogeneous across different M_i values. If the linear model (7.13) is inadequate, transformations of M_i or polynomial terms might be included.

7.3.6 Mixed Effects Hybrid Model

An important assumption in the shared random effects model is that the missingness is independent of the outcome when conditioning on the shared random effects b_i, that is,

$$f(Y_i|M_i, X_i, b_i) = f(Y_i|X_i, b_i).$$

Although this assumption is difficult to test, Yuan and Little (2009) proposed to model $f(Y_i|M_i, X_i, b_i)$ directly. This model is called the mixed effects hybrid model, a combination of the pattern mixture model and the shared random effects model.

Yuan and Little (2009) considered the continuous longitudinal outcome and a monotone missing pattern (dropout). Here, M_i only consists of the dropout time R_i. The outcome process is modeled by a linear mixed model, stratified by the dropout pattern:

$$[Y_i|X_i, R_i = r, b_i] \sim N(X_i\beta^{(r)} + Z_ib_i, \Sigma^{(r)}). \tag{7.15}$$

As discussed in Section 7.3.3.1, if no structure of $\beta^{(r)}$ and $\Sigma^{(r)}$ is assumed, the saturated model (7.15) is not identifiable. A straightforward constraint is to set $\Sigma^{(r)} = \sigma^2 I$, and set $\beta^{(r)}$ to depend linearly on R_i. Note that if $\beta^{(r)} = \beta$ and $\Sigma^{(r)} = \Sigma$, then the model reduces to a shared random effects model.

The dropout process can be modeled by the continuation ratio model, with b_i included as a predictor:

$$\lambda_{ir} \equiv P(R_i = r|R_i \geq r, X_i, b_i) = g(W_{ir}'\theta_1 + b_i'\theta_2),$$

where W_{ir} is expanded by the covariates X_i. Therefore, the dropout probability at time r is given by

$$P(R_i = r|X_i, b_i) = \lambda_{ir} \prod_{k=1}^{r-1}(1 - \lambda_{ik}). \tag{7.16}$$

The joint distribution of R_i and Y_i is factored out as

$$f(R = r, Y_i|X_i, b_i) = f(Y_i|X_i, R_i = r, b_i)P(R_i = r|X_i, b_i),$$

where the first and second terms on the right-hand side are given by (7.15) and (7.16), respectively. The likelihood of the observed data is

$$L(\beta, \Sigma, \theta) = \prod_{i=1}^{N} \int f(Y_i|X_i, R_i = r, b_i)P(R_i = r|X_i, b_i)\mathrm{d}b_i,$$

which is maximized for estimating the model parameters.

When the effect of X_i on the marginal mean Y_i is of interest, one can average the conditional mean over the random effects b_i and the missing pattern R_i:

$$\begin{aligned} E(Y_i|X_i) &= E_{b_i} E_{R_i|b_i}(X_i\beta^{(r)} + Z_i b_i) \\ &= X_i \sum_{k=1}^{K} E_{b_i}\{P(R_i = r|X_i, b_i)\}\beta^{(k)}, \end{aligned}$$

which is a weighted average of $\beta^{(k)}$. As a special case, if the dropout pattern does not depend on the covariates, $P(R_i = r|X_i, b_i)$ can be simply estimated by the proportion of subjects who dropped out at time r.

7.3.7 Comparison of Likelihood-Based Methods

In terms of model interpretation, the selection model and shared random effects model are more natural choices because the regression model of interest $Y_i|X_i$ is directly estimated. The parameters in the pattern mixture model are not of primary interest, and an additional step is often needed to evaluate the relationship of Y_i and X_i unconditional on the missing patterns. For a non-normal outcome variable, this step often leads to a nonlinear covariate effect. In addition, one may argue that the pattern mixture model may not reflect the underlying mechanism that generates the data. For example, if a subject drops out because of being too ill, the outcome process clearly precedes the missingness process. It is less straightforward to assume a distribution of Y_i conditional on the future events R_i.

When model estimation is a major concern, however, the pattern mixture model is more favorable. In most cases, the estimation of the stratified response model can be conducted in standard software packages, such as the SAS GLMMIX procedure. The ''marginal'' missing mechanism model is also easy to estimate, since no missing data are involved. As a comparison, the selection model and the shared random effects model always require the numerical evaluation of intractable integrals. The selection model needs to integrate over the missing outcome, and the shared random effects model needs to integrate over the random effects. Both could be cumbersome as the dimension of integration grows.

For an intermittent missing outcome, the shared random effects model is often easier to use. The missing probability at each observation time can be specified using a generalized linear mixed model. Hence, the nonmonotone missingness barely adds any complexity to the model assumptions. With the pattern mixture model, much more missing data patterns need to be considered: m patterns for dropout compared with 2^m patterns for intermittent missing, where m is the maximum number of observations. Some of the patterns may be sparse, so it would be necessary to collapse the missing patterns. One easy way is to group all the patterns with the same number of missing observations; another way is to consider latent class models. Both methods impose more restrictions on the model. The selection model is generally not recommended for nonmonotone missing data. The missing mechanism model for each time point would involve the current response variable as the predictor. So, the likelihood function involves high-dimensional integration with respect to the missing outcome at every time point.

Finally, we should note that all three types of models make strong and untestable assumptions to identify the parameters. The pattern mixture model borrows information from completers to estimate the conditional mean of the missing subjects. The selection model specifies the exact way that the missing mechanism is influenced by the missing data itself. The shared random effects model assumes that there exists a normally distributed latent variable that affects the outcome process and the missing data process in certain ways. All three models are, in general, not special cases of each other, but rely on different distributional assumptions. From the data analyst's perspective, it is critical that these assumptions are validated based on the scientific knowledge behind the data.

7.4 BAYESIAN ANALYSIS FOR GENERALIZED LINEAR MODELS WITH NONIGNORABLY MISSING COVARIATES

In this section, we discuss the Bayesian analysis for generalized linear models when the covariates are possibly missing nonignorably. Although the Bayesian analysis can be used for other parametric models, Huang et al. (2005) focused on GLMs that are very useful in many applications because the approach can be illustrated more clearly and we can get more specific results of GLMs. Ibrahim et al. (2005) reviewed the approaches for inference in GLMs with ignorable and nonignorable missing data.

Suppose that $\{(\mathbf{x}_1, y_1), \ldots, (\mathbf{x}_n, y_n)\}$ is an independent and identically distributed sample of size n, where each y_i is the response variable and each $\mathbf{x}_i = (x_{i1}, \ldots, x_{ip})'$ is a $p \times 1$ random vector of covariates. The covariates might be missing nonignorably. Let $\mathbf{r}_i = (r_{i1}, \ldots, r_{ip})'$ be a $p \times 1$ missing indicator vector, where $r_{ij} = 1$ if x_{ij} is observed, and 0 otherwise. We also let $\mathbf{x}_{\text{obs},i}$ and $\mathbf{x}_{\text{mis},i}$ be the observed and missing values of $\mathbf{x}_i$. The dimensions of $\mathbf{x}_{\text{obs},i}$ and $\mathbf{x}_{\text{mis},i}$ generally vary with i. Denote D_{obs} as the observed data. The basic task of the Bayesian analysis is to calculate the posterior distribution of parameters of interest. To do that, we need to specify the following distributions:

- the conditional density of y_i given $\mathbf{x}_i$, $f(y_i|\mathbf{x}_i, \beta)$ with parameter β,
- the covariate distribution $f(\mathbf{x}_i|\alpha)$ with parameter α,
- the missingness mechanism $f(\mathbf{r}_i|y_i, \mathbf{x}_i, \phi)$ with parameter ϕ, and
- the joint prior distribution of the parameters (β, α, ϕ), $\pi(\beta, \alpha, \phi)$.

Then, the joint posterior distribution of (β, α, ϕ) and $\pi(\beta, \alpha, \phi|D_{\text{obs}})$ satisfies

$$\pi\left(\beta, \alpha, \phi | D_{\text{obs}}\right) \propto f\left(D_{\text{obs}} | \beta, \alpha, \phi\right) \pi\left(\beta, \alpha, \phi\right), \tag{7.17}$$

where

$$f\left(D_{\text{obs}}|\beta, \alpha, \phi\right) = \int \left\{ \prod_{i=1}^{n} f\left(y_i|\mathbf{x}_i, \beta\right) f\left(\mathbf{x}_i|\alpha\right) f\left(\mathbf{r}_i|y_i, \mathbf{x}_i, \phi\right) d\mathbf{x}_{\text{mis},i} \right\}$$

is the likelihood of the observed data. In the following, we discuss GLMs, the modeling of the covariate distribution and the missingness mechanism, the specification of the prior distribution of (β, α, ϕ), and the computation of the posterior distribution of (β, α, ϕ).

7.4.1 Generalized Linear Models

We specify a GLM for the regression of y_i on $\mathbf{x}_i$ $(i = 1, \ldots, n)$; that is, $(y_i|\mathbf{x}_i)$ has a density in the exponential family with the form

$$f\left(y_i|\mathbf{x}_i, \theta_i, \tau\right) = \exp\left\{a_i^{-1}(\tau)\left[y_i\theta_i - b\left(\theta_i\right)\right] + c\left(y_i, \tau\right)\right\}, \tag{7.18}$$

which is indexed by the canonical parameter θ_i and the scale parameter τ. The function $a_i(\tau)$ is commonly of the form $a_i(\tau) = \tau^{-1}k_i^{-1}$, where k_i is known weight, $i = 1, \ldots, n$. With different settings of functions b and c and parameters θ_i and τ, the GLM (7.18) includes a large class of regression models, such as normal linear regression, logistic and probit regression, Poisson regression, gamma regression, and some proportional hazards models (see McCullagh and Nelder, 1989). Suppose that θ_i

satisfies the equation $\theta_i = \theta(\eta_i)$ with $\eta_i = \mathbf{x}_i^\top \beta$, $i = 1, \ldots, n$, where $\beta = (\beta_1, \ldots, \beta_p)'$ is a $p \times 1$ vector of regression coefficients and θ is a monotone differentiable function. When $\theta_i = \eta_i$, it is called a canonical link. To avoid possible intensive computation, we assume throughout that $\tau = 1$, as in, for example, logistic and Poisson regressions. The methods discussed here can be easily extended to the case in which τ is unknown. Since $\tau = 1$, we write throughout

$$f\left(y_i|\mathbf{x}_i, \theta_i, \tau\right) \equiv f(y_i|\mathbf{x}_i, \beta). \tag{7.19}$$

7.4.2 Modeling the Covariate Distribution

Specification of the covariate distribution is important for the Bayesian analysis when the covariates are possibly missing. Here, we model the joint distribution of $\mathbf{x}_i = (x_{i1}, \ldots, x_{ip})'$ by a sequence of one-dimensional conditional distributions:

$$\begin{aligned} f(x_{i1}, \ldots, x_{ip}|\alpha) &= f(x_{ip}|x_{i1}, \ldots, x_{i,p-1}, \alpha_p) \\ &\quad \times f(x_{i,p-1}|x_{i1}, \ldots, x_{i,p-2}, \alpha_{p-1}) \cdots f(x_{i1}|\alpha_1), \end{aligned} \tag{7.20}$$

where α_j is a vector of parameters for the jth conditional distribution for $j = 1, \cdots, p$, the α_j is distinct, and $\alpha = (\alpha_1, \ldots, \alpha_p)$. It is worth noting that it is not necessary to specify all p conditional distributions. Only the conditional distributions of those covariates with missing values in the n observations should be modeled. Those covariates that are completely observed can be conditioned on. For more strategies in modeling covariate distribution, see Ibrahim et al. (1999) and Chen and Ibrahim (2001).

Modeling the covariate distribution in (7.20) has some advantages. First, it simplifies the prior elicitation for the parameter α. Second, it reduces the computation burden for sampling from the observed data posterior. Specifically, if each of the one-dimensional conditional covariate distributions is in the exponential family and the priors are in the exponential family, then the desired posterior distributions will be log-concave, and the Gibbs sampler via the adaptive algorithm of Gilks and Wild (1992) can be directly applied. Chen and Little (1999) and Chen (2002, 2004) considered nonparametric and semiparametric approaches for modeling the covariate distribution.

7.4.3 Modeling the Missing Data Mechanism

If the missingness of covariates $\mathbf{x}_i$ is independent given y_i and $\mathbf{x}_i$, then the conditional distribution of missingness indicator vector r_i, $f(\mathbf{r}_i|y_i, \mathbf{x}_i, \phi)$, is easily modeled as a product of p independent missing processes. Otherwise, we can model $f(\mathbf{r}_i|y_i, \mathbf{x}_i, \phi)$ using the same idea as factoring out the covariate distribution by a sequence of one-dimensional conditional distributions, that is,

$$p\left(r_{i1},\ldots,r_{ip}|y_i,\mathbf{x}_i,\phi\right)=p\left(r_{ip}|r_{i1},\ldots,r_{i,p-1},y_i,\mathbf{x}_i,\phi_p\right)\\ \times p\left(r_{i,p-1}|r_{i1},\ldots,r_{i,p-2},y_i,\mathbf{x}_i,\phi_{p-1}\right)\cdots p\left(r_{i1}|y_i,\mathbf{x}_i,\phi_1\right), \tag{7.21}$$

where ϕ_k is a vector of parameters for the kth conditional distribution and $\phi = (\phi_1, \ldots, \phi_p)$. Since r_{ik} is binary, it is natural to use a sequence of logistic regressions for (7.21), and hence r_{ik} is log-concave in ϕ_k.

7.4.4 Prior Distributions

It is reasonable to take independent priors for β, α, and ϕ, that is, $\pi(\beta, \alpha, \phi) = \pi(\beta)\pi(\alpha)\pi(\phi)$. In practice, the data analyst often does not have much prior knowledge about the regression parameter of interest, β , or the associations among the covariates, α. For β, we simply take $\pi(\beta) \propto 1$, an improper uniform prior. Note that the covariate distribution is modeled by (7.20) so that we could specify prior distributions for each α_j independently. We could elicit a multivariate normal prior for α_j with large variances. However, we should be more cautious to specify the prior for ϕ.

Huang et al. (2005) showed that when $\pi(\phi)$ is taken to be a uniform improper prior, that is, $\pi(\phi) \propto 1$, the joint posterior distribution $\pi\left(\beta, \alpha, \phi | D_{\text{obs}}\right)$ is improper, that is, $\int f\left(D_{\text{obs}}|\beta,\alpha,\phi\right)\pi\left(\beta,\alpha\right)\mathrm{d}\phi\mathrm{d}\beta\mathrm{d}\alpha = \infty$ for all prior distribution of (β, α), $\pi(\beta, \alpha)$. They also showed that when $\pi(\phi)$ is taken to be proper and $\pi(\beta) \propto 1$, the joint posterior in (7.17) is proper. Therefore, specifying a proper prior for ϕ is an essential and unavoidable task. However, we cannot just specify an arbitrary proper prior for ϕ because inferences about β or α will generally be very sensitive to the choices of the prior for ϕ. So, Huang et al. (2005) proposed a novel class of proper priors for ϕ based on an empirical Bayes approach.

Note that if covariates are all completely observed, then Chen (1985) showed that the posterior distribution of ϕ is asymptotically normal with mean equal to the maximum likelihood estimate of ϕ and covariance matrix equal to the inverse of the negative Hessian matrix. Although some of the covariates might be missing, we can sample the missing values from $f(\mathbf{x}_{\text{mis},i}|\mathbf{x}_{\text{obs},i}, \alpha)$ if α is known. When α is unknown, Huang et al. (2005) simply chose the complete-case estimator of α, $\hat{\alpha}$, which maximizes the likelihood $L(\alpha|\mathbf{x}_1, \ldots, \mathbf{x}_{n_{\text{obs}}})$, where n_{obs} is the number of completely observed subjects, and without loss of generality we assume that the first n_{obs} subjects are completely observed. Then, we independently generate $\mathbf{x}^s_{\text{mis},i} \sim f(\mathbf{x}_{\text{mis},i}|\mathbf{x}_{\text{obs},i}, \hat{\alpha})$ for $s = 1, 2, \ldots, S$ and $i = n_{\text{obs}} + 1, \ldots, n$. For each s, let

$$L^s\left(\phi\right)=\prod_{i=1}^{n_{\text{obs}}} f\left(\mathbf{r}_i|y_i,\mathbf{x}_i,\phi\right)\prod_{i=m+1}^{n} f\left(\mathbf{r}_i|y_i,\mathbf{x}_{\text{obs},i},\mathbf{x}^s_{\text{mis},i},\phi\right)$$

be the "imputed likelihood function" for ϕ. Then, we compute $\hat{\phi}^s$ that maximizes $L^s(\phi)$. Let $I^s\left(\phi\right) = -\partial^2 L^s\left(\phi\right)/\partial\phi\partial\phi^\top$ and $\Phi_0 = \left[S^{-1}\sum_{s=1}^{S} I^s(\hat{\phi}^s)\right]^{-1}$. We generate the missing covariate values S times rather than one time because this yields a much more numerically stable Φ_0. Then, we take the prior for ϕ to be

$$\phi \sim N(\hat{\phi}, c_0\Phi_0), \tag{7.22}$$

where $\hat{\phi} = S^{-1}\sum_{s=1}^{S}\hat{\phi}^s$. Here we introduce a constant c_0 for (7.22) so that we can adjust the prior distribution. But note that as $c_0 \to \infty$, $\pi(\phi)$ converges to an improper uniform prior, which could result in an improper posterior distribution. Therefore, we might not set a large value for c_0, and it is necessary to conduct a detailed sensitivity analysis on the posterior distribution with respect to the choice of c_0.

7.4.5 Posterior Computation

Now we consider computation of joint posterior distribution $\pi(\beta, \alpha, \phi | D_{\text{obs}})$ given by (7.17). A standard Gibbs sampler requires sampling from the following full conditional distributions: (i) $[\beta | X_{\text{mis}}, D_{\text{obs}}]$, (ii) $[\alpha | X_{\text{mis}}, D_{\text{obs}}]$, (iii) $[\phi | X_{\text{mis}}, D_{\text{obs}}]$, and (iv) $[X_{\text{mis}} | \beta, \alpha, \phi, D_{\text{obs}}]$, where X_{mis} denotes the vector of all missing covariates. This might not be an easy task because it may not be possible to obtain closed forms of the above four full conditional posterior distributions for the GLM and the missing data mechanism (7.21). However, the adaptive rejection algorithm (Gilks and Wild, 1992) may be readily applied, as long as the above four conditional posterior distributions are log-concave in each component of $(\beta, \alpha, \phi, X_{\text{mis}})$. To achieve this, we can just let the joint prior distribution $\pi(\beta, \alpha, \phi)$ be log-concave in each component of (β, α, ϕ) and $f(\mathbf{x}_i | \alpha)$ log-concave in each component of $\mathbf{x}_i$.

Even though the joint posterior is proper when $\pi(\phi)$ is proper, the mixing of the Gibbs sampling algorithm is slow when $\pi(\phi)$ is chosen to be proper but relatively flat. To remedy this, Albert and Chib (1993) fitted a Bayesian nonignorable model with the probit link for the missing data mechanism $\mathbf{r}_i$ in (7.21) and introduced a vector of latent variables ξ_i for each $\mathbf{r}_i$ proposed by Albert and Chib (1993). Then, the collapsed Gibbs sampler is used. That is, instead of sampling from the above four full conditional distributions, we sample from the following full conditional distributions: (i) $[\beta | X_{\text{mis}}, D_{\text{obs}}]$, (ii) $[\alpha | X_{\text{mis}}, D_{\text{obs}}]$, (iii) $[\phi | \xi, X_{\text{mis}}, D_{\text{obs}}]$, (iv) $[X_{\text{mis}} | \beta, \alpha, \xi, D_{\text{obs}}]$, and (v) $[\xi | \phi, X_{\text{mis}}, D_{\text{obs}}]$, where $\xi = (\xi_1^{\top}, \ldots, \xi_n^{\top})'$. It is easy to sample from $[\phi | \xi, X_{\text{mis}}, D_{\text{obs}}]$ and $[\xi | \phi, X_{\text{mis}}, D_{\text{obs}}]$, since the former is a multivariate normal distribution and the latter is the product of univariate truncated normal. Next, we draw X_{mis} from $[X_{\text{mis}} | \beta, \alpha, \xi, D_{\text{obs}}]$ after collapsing on ϕ. Due to the probit link and the introduction of the latent variable ξ, we are able to analytically integrate out ϕ. Furthermore, it is worth noting that we can extend the probit link to a much more general scale mixture of normal links.

7.5 MULTIPLE IMPUTATION

Imputation is a common and practical method for dealing with missing data where missing values are replaced with plausible guesses. While single imputation cannot reflect the sampling uncertainty, multiple imputation is generally recommended. Denote by γ the parameters of interest related to the joint distribution of $(\mathbf{X}, Y)$, with $\mathbf{X}$ the covariate vector and Y the response variable. The basic idea behind MI is as follows:

- Create M "complete" data sets by filling the missing values with a random draw from the posterior predictive distribution.
- Obtain parameter estimates $\hat{\gamma}^{(m)}$ for the mth imputed data set, $m = 1, \ldots, M$.
- The estimator of γ is $\hat{\gamma} = M^{-1} \sum_{m=1}^{M} \hat{\gamma}^{(m)}$.
- The variance estimate of $\hat{\gamma}$, $\hat{V}_{\mathrm{MI}}$, is given by Rubin's rule, that is, $\hat{V}_{\mathrm{MI}} = \bar{V} + \left(1 + M^{-1}\right) \hat{B}$. Here, $\bar{V}$ is the average within-imputation variance given by $\bar{V} = M^{-1} \sum_{m=1}^{M} \hat{V}^{(m)}$, with $\hat{V}^{(m)}$ being the variance estimate from the mth imputed data set, obtained by assuming that the imputed data set is the complete data set. $\hat{B} = (M-1)^{-1} \sum_{m=1}^{M} (\hat{\gamma}^{(m)} - \hat{\gamma}) \times (\hat{\gamma}^{(m)} - \hat{\gamma})'$ is the between-imputation variance.

In Section 7.5.1, we introduce the proper multiple imputation by the Bayesian analysis and the approximate Bayesian bootstrap (ABB) hot deck imputation.

7.5.1 Proper Multiple Imputation by Bayesian Analysis

To illustrate the basic idea of proper multiple imputation, we use the models considered in Section 7.4 as an example. Recall that β, α, and ϕ are parameters related to the regression, covariate distribution, and missing data mechanism, respectively. Also recall that $\mathbf{x}_{\mathrm{obs},i}$ and $\mathbf{x}_{\mathrm{mis},i}$ denote the observed and missing components of $\mathbf{x}_i$. Then, a proper MI specifies a prior distribution for (β, α, ϕ) and imputes $\mathbf{x}_{\mathrm{mis},i}$ from its posterior predictive distribution, given by

$$
\begin{aligned}
& f\left(\mathbf{x}_{\mathrm{mis},i} | y_i, \mathbf{x}_{\mathrm{obs},i}\right) \\
& \quad \propto \iiint f\left(\mathbf{x}_{\mathrm{mis},i} | y_i, \mathbf{x}_{\mathrm{obs},i}, \mathbf{r}_i, \beta, \alpha, \phi\right) \pi\left(\beta, \alpha, \phi | D_{\mathrm{obs}}\right) \mathrm{d}\beta \mathrm{d}\alpha \mathrm{d}\phi,
\end{aligned}
$$

where $\pi\left(\beta, \alpha, \phi | D_{\mathrm{obs}}\right)$ is the joint posterior distribution for (β, α, ϕ) given the observed data. Here,

$$
f\left(\mathbf{x}_{\mathrm{mis},i} | y_i, \mathbf{x}_{\mathrm{obs},i}, \mathbf{r}_i, \beta, \alpha, \phi\right) \propto f\left(y_i | \mathbf{x}_i, \beta\right) f\left(\mathbf{x}_i | \alpha\right) f\left(\mathbf{r}_i | y_i, \mathbf{x}_i, \phi\right),
$$

which can be sampled via the Gibbs sampler. As indicated in Section 7.4, a proper prior for ϕ is required to guarantee a proper posterior distribution $\pi\left(\beta, \alpha, \phi | D_{\mathrm{obs}}\right)$. The computation of $\pi\left(\beta, \alpha, \phi | D_{\mathrm{obs}}\right)$ can be seen in Section 7.4.

7.5.2 Approximate Bayesian Bootstrap Hot Deck Imputation

Hot deck imputation is a widely used imputation technique for missing data problems in practice (Andrige and Little, 2010). In hot deck imputation, the missing values of one or more variables of a subject (called the donee or recipient) are replaced with observed values from a ''similar'' subject (called the donor). Many versions of the hot deck imputation have been developed, for example, the nearest-neighbor imputation and its variations. There are some advantages of the hot deck imputation. Since imputed values are from other observed values, the imputations seem realistic and they are not outside the range of the variables. Another advantage is that it does not require an explicit model for the distribution of the missing values.

Usually the hot deck imputation cannot reveal parameter uncertainty. To overcome this, Siddique and Belin (2008a, 2008b) incorporated an approximate Bayesian bootstrap (Rubin and Schenker, 1986; Demirtas et al., 2007) into hot deck imputation. The procedure is given as follows:

- Use the ABB to generate a random donor pool.
- Use hot deck imputation where the donors are selected from the donor pool generated by the last step.
- Repeat the above steps M times and get M ''complete'' data sets. Then estimate the parameter(s) of interest as multiple imputation methods do, and the variance of the estimator is estimated by Rubin's rule.

In the following sections, we introduce how to use the ABB to generate a random donor pool for nonignorable missingness, and how to implement the hot deck imputation. To illustrate the ABB hot deck imputation clearly, we consider the case where there is a single variable with missing values, which is denoted by Y. The set of values of Y that are observed is denoted by $\mathcal{Y}_{\text{obs}}$, and the size of the set is denoted by n_{obs}. When more variables have missing values, iterative imputation can be used. The idea is similar to the multiple imputation with chained equations. First, use mean imputation or random draws from the observed values as starting values for the missing values. Then reimpute the first variable in the data set by treating the other variables as completely observed. Similarly, the other variables are reimputed. This procedure is iterated until convergence. Siddique and Belin (2008b) introduced some diagnostic statistics for the convergence. They also showed in one example that the estimates by 10 iterations of the hot deck procedure were not substantially different from estimates resulting from iterating the procedure until convergence diagnostics were satisfied.

7.5.2.1 A Nonignorable ABB and Its Variations The ABB is proposed by Rubin and Schenker (1986) for generating multiple imputations when the missing mechanism is ignorable. For each of M replications, n_{obs} cases are drawn from $\mathcal{Y}_{\text{obs}}$ randomly with replacement (i.e., with equal probability). To deal with the nonignorable missing data, Rubin and Schenker (1991) proposed to draw n_{obs} cases from $\mathcal{Y}_{\text{obs}}$ and the probabilities of selection of subjects are related with their observed Y values. Specifically, without loss of generality, we assume that $y_1, \ldots, y_{n_{\text{obs}}}$ are included in $\mathcal{Y}_{\text{obs}}$. The probability of selection for $y_i \in \mathcal{Y}_{\text{obs}}$ defined by Rubin and Schenker (1991) is

$$\frac{y_i^c}{\sum_{j=1}^{n_{\text{obs}}} y_j^c} \tag{7.23}$$

for some c. Suppose every $y_j > 0$. If subjects with larger values of Y are conceivably more likely to be missing, then we could choose a positive c; otherwise, choose a negative c. The ABBs where the probability of selection is defined by (7.23) with $c = -1, 1, 2$, and 3 are referred to as ''inverse-to-size ABB,'' ''proportional-to-size ABB,'' ''proportional-to-size-squared ABB,'' and ''proportional-to-size-cubed ABB,'' respectively. When $c = 0$, this is an ignorable ABB.

If $y_i \leq 0$, then the selection probability defined by (7.23) might be less than or equal to 0. To ensure that the selection probabilities are positive, we need to make some modifications for (7.23). Meanwhile, we may want to keep the property that the selection probabilities preserve the order of y_i^c, that is, for $y_i, y_j \in \mathcal{Y}_{\text{obs}}$, if $y_i > y_j$, then the selection probability of y_i is larger than that of y_j. Taking these issues into account, we can redefine the selection probability for $y_i \in \mathcal{Y}_{\text{obs}}$ as

$$\frac{(y_i + |\alpha| + |\alpha - \beta|)^c}{\sum_{j=1}^{n_{\text{obs}}}(y_j + |\alpha| + |\alpha - \beta|)^c},$$

where α and β are the smallest and second smallest values of $\mathcal{Y}_{\text{obs}}$, respectively.

Siddique and Belin (2008a) considered a number of variations on the idea of Rubin and Schenker (1991). For example, the probability of selection of $y_i \in \mathcal{Y}_{\text{obs}}$ is proportional to $|y_i - Q_p(\mathcal{Y}_{\text{obs}})|^c$, where the notation $Q_p(\mathcal{Y}_{\text{obs}})$ represents the pth quantile of $\mathcal{Y}_{\text{obs}}$. When $p = 2$, $Q_p(\mathcal{Y}_{\text{obs}})$ is the median of the observed Y values and the approach is referred to as a ''U-shaped ABB,'' because when $c > 0$, observations in the extremes of the distribution of $\mathcal{Y}_{\text{obs}}$ have a greater chance to be selected into the donor pool than the observations that are close to the median. When $p = 1$, we refer to the corresponding ABB as a ''Fishhook ABB.''

Nonignorable ABBs account for the nonignorability of missingness. Different ABBs correspond to different missing data assumptions, which cannot be verified by data. Relevant subject matter experts might be able to answer which ABBs are plausible, and a sensitivity analysis is required. In practice, it is suggested that one analyze data using several plausible ABBs and then explore the sensitivity of inferences to different missing data assumptions.

7.5.2.2 Hot Deck Imputation The nonignorable ABBs provide us a random donor pool, which we denote by $\mathcal{Y}_{\text{obs}}^*$, where the number of y_i is w_i, $i = 1, \ldots, n_{\text{obs}}$. Then, how can we select a donor for a donee from the donor pool $\mathcal{Y}_{\text{obs}}^*$? Siddique and Belin (2008a, 2008b) suggested selecting donors with probability inversely proportional to their distance from the donee. Let D_{i0} be the distance between donee 0 and donor i. Then, donor for donee 0 is selected from the distribution where the selection probability of donor i is

$$l_{0i} = \frac{(1/D_{0i})w_i}{\sum_{j=1}^{n_{\text{obs}}}(1/D_{0j})w_j},$$

which ensures that $\sum_{i=1}^{n_{\text{obs}}} l_{0i} = 1$. The expected value of the imputation for donee 0 is

$$E(y_0|\mathcal{Y}^*_{\text{obs}}, l_{01}, \ldots, l_{0,n_{\text{obs}}}) = \sum_{i=1}^{n_{\text{obs}}} l_{0i}y_i.$$

We can see that unlike the other hot deck imputation, this approach allows all donors to be eligible and have a nonzero selection probability, but donors closest to the donee have the greatest selection probability.

Then the question arises of how to define the distance D_{0i}. Predictive mean matching (Little, 1988; Schenker and Taylor, 1996) is a popular way to define the distance and is employed by Siddique and Belin (2008a, 2008b). In predictive mean matching, Y has one or more covariates and the regression parameters are estimated based on the donor pool. Using the estimated regression parameters, predicted values are calculated for all subjects. Then, the distance between donee 0 and donor i is defined based on their predicted values. Specifically, let $\mathbf{Y}_{\text{obs}} = (y_1, \ldots, y_{n_{\text{obs}}})'$, X the corresponding covariate matrix with row vector $x_i^\top$, and x_0 the covariate vector of donee 0. Then, $\hat{y}_i = x_i\hat{B}$ and $\hat{y}_0 = x_0\hat{B}$, where $\hat{B} = (X^\top WX)^{-1}X^\top W\mathbf{Y}_{\text{obs}}$. Then, the distance D_{0i} is defined by

$$D_{0i} = \left(|\hat{y}_0 - \hat{y}_i| + \delta\right)^k, \tag{7.24}$$

where

$$\delta = \min|\hat{y}_0 - \hat{y}_j|, \quad \text{for all } j = 1, \ldots, n_{\text{obs}}, \text{ with } \hat{y}_0 \neq \hat{y}_j,$$

is a positive offset to avoid zero distance and k is a given constant. If there are no two subjects i and j with $\hat{y}_i = \hat{y}_j$, then δ can be neglected.

The exponent k in (7.24) is a closeness parameter to adjust the selection probabilities of the donors. As $k \to \infty$, this procedure reduces to a nearest-neighbor imputation where the donor closest to the donee is selected with probability 1. When k equals 0, each donor has equal probability of selection, which is equivalent to an ignorable hot deck. In practice, one usually chooses a positive value for k. Siddique and Belin (2008a) considered an example where a closeness parameter value of around 3 appeared to be reasonable to favor nearby donors while allowing the selection probabilities to decline smoothly as a function of distance.

It is worth noting that, just like the role of the bandwidth in the kernel regression, the closeness parameter also has an impact on the bias and variance of the imputed values. A smaller value of the closeness parameter indicates a lower variance (averaging over more donors) and a higher bias (since in nonignorable missingness, donors are not equally good matches).

7.6 INVERSE PROBABILITY WEIGHTED METHODS

In Chapters 4, 5 and 6, we discussed the use of the inverse probability weighted method and the augmented inverse probability weighted method to handle ignorable missing data. Here, we describe how the inverse probability weighted methods apply to nonignorable missing data. Rotnitzky et al. (1998) discussed semiparametric regression for repeated outcomes with nonignorable nonresponse, which is introduced in Section 7.6.1. In Section 7.6.2, we discuss the estimation of the marginal mean of outcome measured at a fixed time; see Scharfstein et al. (1999) for more details.

7.6.1 Semiparametric Regression for Repeated Outcomes

7.6.1.1 Model and Notation Consider a longitudinal study conducted over a prespecified interval from time 0 to T. Let $\mathbf{Y}_i = \left(Y_{i1}, \ldots, Y_{iT}\right)'$ be the outcome vector of the ith subject, where Y_{it} is the outcome of the ith subject measured at time t, $t = 1, \ldots, T$, and $i = 1, \ldots, n$. Let Y_{i0} be the baseline outcome if it is measured. Let X_{i0} denote the baseline covariate vector of the ith subject measured prior to the start of follow-up. To allow for time-dependent covariates, we let X_{it} be the covariate vector measured at time t and $\mathbf{X}_i = \left(X_{i0}^{\top}, \ldots, X_{iT}^{\top}\right)'$. The interest is to study the association between the outcome and covariates.

The mean regression model of Y_{it} on $\mathbf{X}_i$ is given by

$$E\left(Y_{it}|\mathbf{X}_i\right) = g_t\left(\mathbf{X}_i; \beta^*\right), \quad t = 1, \ldots, T, \tag{7.25}$$

where β^* is a $p \times 1$ vector of unknown parameters of interest. Here, we assume that for each $\mathbf{X}$, $g_t\left(\mathbf{X}; \beta\right)$ are known smooth functions of β, which could be linear or nonlinear functions of β. With different choices of $g_t(\cdot;\cdot)$, model (7.25) allows for the dependence of the mean of Y_{it} on baseline covariates, deterministic functions of time, interactions between time and baseline covariates, and functions of covariate history up to time t.

Although our interest is in the conditional mean regression of Y_{it} on $\mathbf{X}_i$, we may also observe a vector of time-dependent covariates V_{it}, which is of no substantive interest. We denote $W_{it} = \left(V_{it}^{\top}, Y_{it}\right)', t = 1, \ldots, T$; $W_{i0} = \left(X_{i0}^{\top}, V_{i0}^{\top}, Y_{i0}\right)'$; and $\mathbf{W}_i = \left(W_{i0}^{\top}, \ldots, W_{iT}^{\top}\right)'$.

Robins (1992, 1994) and Robins, Greenland and Hu (1999) showed that regression models cannot be used for analyzing the effect of a nonancillary covariate process. We assume that $\mathbf{X}_i$ is an ancillary covariate process; that is, given $\left(X_{i0}^{\top}, \ldots, X_{it}^{\top}\right)$,

$X_{i(t+1)}$ is conditionally independent of $\left(W_{i0}^{\top}, \ldots, W_{it}^{\top}\right)$. Then we can obtain

$$E\left(Y_{it}|\mathbf{X}_i\right) = E\left(Y_{it}|X_{i0}, \ldots, X_{it}\right).$$

This implies that the conditional mean of the outcome at time t depends on the covariates measured up to time t.

We assume that W_{i0} and $\mathbf{X}_i$ are always observed, but Y_{it} and V_{it} are either both observed or both missing for $t = 1, \ldots, T$. Let R_{it} denote the missing indicator for time t, $t = 0, \ldots, T$; that is, $R_{it} = 1$ if W_{it} is observed and $R_{it} = 0$ otherwise. Denote $\mathbf{R}_i = \left(R_{i1}, \ldots, R_{iT}\right)'$ that takes values in the set $\left\{\mathbf{r} = \left(r_1, \ldots, r_T\right)' : r_t = 0 \text{ or } 1, 1 \leq t \leq T\right\}$. Define $\pi_i(\mathbf{r}) = P\left(\mathbf{R}_i = \mathbf{r}|\mathbf{W}_i\right)$. We assume that the response probabilities are characterized by a $q \times 1$ vector of parameter α^*, that is, $\pi_i(\mathbf{r}) \equiv \pi_i(\mathbf{r}; \alpha^*)$, which is constructed similarly to (7.21). Let $\bar{\mathbf{R}}_{it} = \left(R_{i1}, \ldots, R_{i(t-1)}\right)'$ and, for notational convenience, we define $\bar{R}_{i1} = 1$. By letting $\lambda_{it} = P\left(R_{it} = 1|\bar{\mathbf{R}}_{it}, \mathbf{W}_i\right)$, we can factor $\pi_i(\mathbf{r})$ as

$$\pi_i(\mathbf{r}) = \prod_{t=1}^{T} \lambda_{it}^{r_t}\{1 - \lambda_{it}\}^{1-r_t}, \tag{7.26}$$

where λ_{it} is modeled as

$$\text{logit } \lambda_{it} = h_t\left(\bar{\mathbf{R}}_{it}, \mathbf{W}_i; \alpha^*\right). \tag{7.27}$$

Here, $h_t(\cdot, \cdot; \cdot)$, $t = 1, \ldots, T$, is a known function. Although λ_{it} is parameterized by its logistic transformation, its functional form is arbitrary because h_t can take any form. It is worth noting that from (7.27) the conditional probabilities of response at time t may depend on the possibly unobserved past and current values of V_{ij} and Y_{ij}, $j \leq t$, as well as future values of these variables. To ensure that we have enough number of subjects that are completely observed, we assume $\pi_i(\mathbf{1})$ is bounded away from 0 with probability 1; that is,

$$\pi_i(1) > \sigma > 0, \text{ with probability } 1, \tag{7.28}$$

where $\mathbf{1}$ is a vector of 1's.

Let $\mathbf{W}_{(\mathbf{r})i}$ denote the vector formed by the observed part of $\mathbf{W}_i$ when $\mathbf{R}_i = \mathbf{r}$. For example, if $\mathbf{0}$ denotes the $T \times 1$ vector of 0's, then $\mathbf{W}_{(\mathbf{0})i} = W_{i0}$, and if $\mathbf{r} = (1, 1, 0, \ldots, 0)$, then $\mathbf{W}_{(\mathbf{r})i} = \left(W_{i0}^{\top}, W_{i1}^{\top}, W_{i2}^{\top}\right)'$. Robins and Rotnitzky (1995) and Robins et al. (1995) described methods for making inferences about β^* in the conditional mean model (7.25) when nonresponse is ignorable and parameterized, that is,

$$P\left(\mathbf{R}_i = \mathbf{r}|\mathbf{W}_i\right) = P\left(\mathbf{R}_i = \mathbf{r}|\mathbf{W}_{(\mathbf{r})i}\right), \tag{7.29}$$

and $P\left(\mathbf{R}_i = \mathbf{r}|\mathbf{W}_{(\mathbf{r})i}\right)$ is known up to a vector of parameters. When (7.29) does not hold, the nonresponse process is nonignorable. In the following section, we describe numerically simple estimators that are consistent and asymptotically normal for the parameter β^* for the semiparametric conditional mean model (7.25), when the response probabilities follow the parametric model (7.27), condition (7.28) holds, and nonresponse is nonignorable. We first discuss the estimation when the probabilities of response are known, or equivalently α^* is known, and then extend it to the cases where α^* is unknown.

7.6.1.2 Estimation When the Probabilities of Response Are Known

The inverse probability weighted method or the augmented inverse probability weighted method can be used to handle the data with nonignorable nonresponse if the response probabilities are known or can be estimated consistently. In almost all of the practical cases where there is nonignorable nonresponse, the response probabilities are unknown to the data analyst. However, to illustrate the methods, we first assume that the response probabilities $\pi_i(\mathbf{r}) = P\left(\mathbf{R}_i = \mathbf{r} | \mathbf{W}_i\right)$ are known. We discuss the cases where the response probabilities are unknown in the next section.

When response probabilities are known, we estimate β^* by solving the following estimating equation:

$$\sum_{i=1}^{n} \frac{I\left(\mathbf{R}_i = \mathbf{1}\right)}{\pi_i\left(\mathbf{1}\right)} \mathbf{d}\left(\mathbf{X}_i; \beta\right) \left\{\mathbf{Y}_i - \mathbf{g}\left(\mathbf{X}_i; \beta\right)\right\} = 0, \tag{7.30}$$

with respect to β, where $\mathbf{d}\left(\mathbf{X}_i; \beta\right)$ is a given $p \times T$ matrix of functions of $\mathbf{X}_i$ and β, and $\mathbf{g}\left(X_i; \beta\right) = \left(g_1\left(\mathbf{X}_i; \beta\right), \ldots, g_T\left(\mathbf{X}_i; \beta\right)\right)'$. The solution to (7.30) is the so-called inverse probability weighted estimator of β^*. Note that the estimating function (7.30) is unbiased, that is,

$$\begin{aligned} &E\left[E\left\{\frac{I\left(\mathbf{R}_i = \mathbf{1}\right)}{\pi_i\left(\mathbf{1}\right)} \mathbf{d}\left(\mathbf{X}_i; \beta\right) \left\{\mathbf{Y}_i - \mathbf{g}\left(\mathbf{X}_i; \beta^*\right)\right\} | \mathbf{W}_i\right\}\right] \\ &= E\left[\mathbf{d}\left(\mathbf{X}_i; \beta\right) \left\{\mathbf{Y}_i - \mathbf{g}\left(\mathbf{X}_i; \beta^*\right)\right\}\right] \\ &= 0. \end{aligned}$$

Under mild regularity conditions, we can show that the inverse probability weighted estimator is a consistent and asymptotically normal estimator of β^*.

Note that only the subjects that are completely measured at all times contribute to (7.30). The inverse probability weighted estimation does not effectively use all available data. To increase estimation efficiency, we now consider the augmented inverse probability weighted estimation, which solves the following class of estimating equation:

$$\sum_{i=1}^{n} \left[\frac{I\left(\mathbf{R}_i = \mathbf{1}\right)}{\pi_i\left(\mathbf{1}\right)} \mathbf{d}\left(\mathbf{X}_i; \beta\right) \left\{\mathbf{Y}_i - \mathbf{g}\left(\mathbf{X}_i; \beta\right)\right\} + A_i\right] = 0 \tag{7.31}$$

with augmentation

$$A_i = \sum_{\mathbf{r}\neq\mathbf{1}} \left\{ I\left(\mathbf{R}_i = \mathbf{r}\right) - \frac{I\left(\mathbf{R}_i = \mathbf{1}\right)}{\pi_i\left(\mathbf{1}\right)} \pi_i\left(\mathbf{r}\right) \right\} \phi_{\mathbf{r}}\left(\mathbf{W}_{(\mathbf{r})i}\right). \tag{7.32}$$

Note that here $\phi_{\mathbf{r}}\left(\mathbf{W}_{(\mathbf{r})i}\right)$ can be chosen as an arbitrary $p \times 1$ vector function of the data $\mathbf{W}_{(\mathbf{r})i}$ that are observed when $\mathbf{R}_i = \mathbf{r}$. Also note from (7.32) that we need to calculate $\pi_i(\mathbf{r})$ only when the ith subject is completely observed. Thus, A_i is a function of the observed data that is computable. It can be seen from (7.31) that subjects with incomplete observations contribute to the estimation of β^* through the term A_i. The consistency and asymptotic normality of the augmented inverse probability weighted estimator of β^* can be obtained by noting that for each $\mathbf{r}$,

$$\begin{aligned} &E\left[\sum_{r\neq 1} \left\{ I\left(\mathbf{R}_i = \mathbf{r}\right) - \frac{I\left(\mathbf{R}_i = \mathbf{1}\right)}{\pi_i\left(\mathbf{1}\right)} \pi_i\left(\mathbf{r}\right) \right\} \phi_{\mathbf{r}}\left(\mathbf{W}_{(\mathbf{r})i}\right) | \mathbf{W}_i \right] \\ &= \left\{ \pi_i\left(\mathbf{r}\right) - \frac{\pi_i\left(\mathbf{1}\right)}{\pi_i\left(\mathbf{1}\right)} \pi_i\left(\mathbf{r}\right) \right\} \phi_{\mathbf{r}}\left(\mathbf{W}_{(\mathbf{r})i}\right) = 0. \end{aligned} \tag{7.33}$$

7.6.1.3 Estimation When the Probabilities of Response Are Unknown

Now we consider estimation of β^* when the response probabilities are unknown. Two different techniques can be used when the response probabilities are unknown. The first one is to estimate the response probabilities consistently and use them as if they were known. The second one is to estimate the response probabilities and β^* simultaneously.

We have assumed a parametric model for the conditional response probabilities $\pi_i(\mathbf{r})$ defined by (7.26) and (7.27) with a $q \times 1$ parameter vector α^*, that is, $\pi_i(\mathbf{r}) = \pi_i(\mathbf{r}; \alpha^*)$. When nonresponse is ignorable, α^* could be estimated by maximizing the likelihood $\prod_{i=1}^n \pi_i(\mathbf{R}_i; \alpha)$ with respect to α, where

$$\pi_i\left(\mathbf{r}; \alpha\right) = \prod_{t=1}^T \left\{\lambda_{it}\left(\alpha\right)\right\}^{r_t} \left\{1 - \lambda_{it}\left(\alpha\right)\right\}^{1-r_t}.$$

However, when the nonresponse is nonignorable, $\pi_i(\mathbf{r}; \alpha)$ involves the unobserved values, and thus $\pi_i(\mathbf{r}; \alpha)$ is not computable except for $\mathbf{r} = \mathbf{1}$. So, we need another estimation method. Note from (7.33) that with true parameter α^*, the augmentation term A_i is unbiased, that is, $E(A_i) = 0$. This helps us develop a method for estimating α^*. A consistent and asymptotically normal estimator of α^* can be obtained by solving the following estimating equation:

$$\sum_i A_i(\alpha) = 0, \tag{7.34}$$

where

$$A_i(\alpha) = \sum_{\mathbf{r} \neq \mathbf{1}} \left\{ I\left(\mathbf{R}_i = \mathbf{r}\right) - \frac{I\left(\mathbf{R}_i = \mathbf{1}\right)}{\pi_i\left(\mathbf{1}; \alpha\right)} \pi_i\left(\mathbf{r}; \alpha\right) \right\} \phi_{\mathbf{r}}\left(\mathbf{W}_{(\mathbf{r})i}\right). \tag{7.35}$$

Now we can estimate β^* by solving (7.31) with $\pi_i(\mathbf{r})$ replaced by $\pi_i(\mathbf{r}; \hat{\alpha})$, where $\hat{\alpha}$ is the solution to (7.34). However, this is a special case, which can be included in a class of methods jointly estimating β^* and α^*. We define the joint estimating equations of β^* and α^* by

$$\sum U_i(\beta; \alpha; d, \phi) = 0, \tag{7.36}$$

where

$$U_i\left(\beta; \alpha; d, \phi\right) = \frac{I\left(\mathbf{R}_i = \mathbf{1}\right)}{\pi_i\left(\mathbf{1}; \alpha\right)} \begin{bmatrix} \mathbf{d}^{(1)}\left(\mathbf{X}_i; \beta\right) \\ \mathbf{d}^{(2)}\left(\mathbf{X}_i; \beta\right) \end{bmatrix} \left\{\mathbf{Y}_i - \mathbf{g}\left(\mathbf{X}_i; \beta\right)\right\} - \begin{bmatrix} A_i^{(1)}\left(\alpha\right) \\ A_i^{(1)}\left(\alpha\right) \end{bmatrix}.$$

Here, $\mathbf{d}^{(1)}\left(\mathbf{X}_i; \beta\right)$ and $\mathbf{d}^{(2)}\left(\mathbf{X}_i; \beta\right)$ are $p \times T$ and $q \times T$ functions and $A_i^{(1)}(\alpha)$ and $A_i^{(2)}(\alpha)$ are defined as in (7.35), with $\phi_{\mathbf{r}}\left(\mathbf{W}_{(\mathbf{r})i}\right)$ replaced by $p \times 1$ and $q \times 1$ functions $\phi_{\mathbf{r}}^{(1)}\left(\mathbf{W}_{(\mathbf{r})i}\right)$ and $\phi_{\mathbf{r}}^{(2)}\left(\mathbf{W}_{(\mathbf{r})i}\right)$. With different choices of the functions $\mathbf{d}^{(j)}\left(\mathbf{X}_i; \beta\right)$ and $\phi_{\mathbf{r}}^{(j)}\left(\mathbf{W}_{(\mathbf{r})i}\right)$, $j = 1, 2$, the solutions to (7.36) essentially contain all regular asymptotic linear estimators of β^*; that is, any RAL (regular and asymptotically linear) estimator of β^* is asymptotically equivalent to an estimator by solving (7.36) with some functions $\mathbf{d}^{(j)}\left(\mathbf{X}_i; \beta\right)$ and $\phi_{\mathbf{r}}^{(j)}\left(\mathbf{W}_{(\mathbf{r})i}\right)$, $j = 1, 2$. Specially, if we set $\mathbf{d}^{(2)}\left(\mathbf{X}_i; \beta\right) = 0$, then the resulting estimation is equivalent to estimating α^* from the solution of (7.34) and in turn estimating β^* from (7.31) using $\pi_i(\mathbf{r}; \hat{\alpha})$.

As discussed by Rotnitzky et al. (1998), the simultaneous estimating equation (7.36) guarantees root-n consistent estimation of β^* and α^* and improves the efficiency. Denote $\gamma^* = \left(\beta^{*\top}, \alpha^{*\top}\right)'$. Rotnitzky et al. (1998) showed that under mild regularity conditions, with probability approaching 1, there exists a unique solution $\hat{\gamma} = \left(\hat{\beta}^\top, \hat{\alpha}^\top\right)'$ to (7.36), and $\sqrt{n}(\hat{\gamma} - \gamma^*)$ is asymptotically normal with mean 0 and variance $\Upsilon = \Gamma^{-1}\Omega\left(\Gamma^{-1}\right)'$, where $\Gamma = E\left\{\partial U_i\left(\beta^*; \alpha^*; d, \phi\right) / \partial\left(\beta, \alpha\right)\right\}$ and $\Omega = \mathrm{var}\left\{U_i\left(\beta^*; \alpha^*; d, \phi\right)\right\}$. It can be seen that Υ can be consistently estimated by $\hat{\Upsilon} = \hat{\Gamma}^{-1}\hat{\Omega}\left(\hat{\Gamma}^{-1}\right)'$, where $\hat{\Gamma} = n^{-1} \sum \partial U_i\left(\beta^*; \alpha^*; d, \phi\right) / \partial\left(\beta, \alpha\right)$ and $\hat{\Omega} = n^{-1} \sum U_i\left(\beta^*; \alpha^*; d, \phi\right) U_i^\top\left(\beta^*; \alpha^*; d, \phi\right)$. The asymptotic variance of $\hat{\beta}$ and its estimator are the upper leftmost $p \times p$ matrices of Υ and $\hat{\Upsilon}$, respectively.

7.6.2 Estimation of Marginal Mean of an Outcome from Longitudinal Data

7.6.2.1 Models and Identification In Section 7.6.1, we were interested in the inference on the conditional expectation of an outcome Y at some follow-up time of a longitudinal study. In this section, we discuss the inference on the marginal mean of an outcome Y measured at the end of the follow-up period, denoted by T. Some subjects may drop out of the study nonignorably at any time before time T. The dropout time can be modeled as a survival outcome, but the difficulty is that the longitudinal process of Y and the drop-out process are correlated. Additional settings about the longitudinal study are given below.

Let Q denote the time to drop out. Then Y is observed if and only if $Q \geq T$. Denote $\Delta = I(Q \geq T)$ as the dropout indicator. In addition to Y, there may be other time-dependent variables V recorded until the subject drops out, which are denoted by $\bar{V}(Q)$. Here, $\bar{V}(Q) = \{V(t) : 0 \leq t \leq Q\}$ and $V(t)$ is the observed value of V at time t. Nonignorable dropout means that dropout time may depend on $(\bar{V}(T), Y)$ that would be observed in the absence of dropout. However, here we only consider the cases where dropout time depends on the possibly unobserved future only through Y. For notational convenience, for subjects who do not drop out, we set the dropout time Q equal to the end of follow-up time T. The data that we observe are n i.i.d. copies of $O = \left(Q, \Delta, \Delta Y, \bar{V}(Q)\right)$:

$$\left\{O_i = \left(Q_i, \Delta_i, \Delta_i Y_i, \bar{V}_i(Q_i)\right) : i = 1, \ldots, n\right\}.$$

In the following, we introduce two semiparametric proportional hazards models for the dropout time Q on Y and $\bar{V}(Q)$, and discuss their identifiability. Denote by $\lambda_Q\left(t|\bar{V}(T), Y\right)$ the conditional hazard of Q given the data $(\bar{V}(T), Y)$, that is, $\lambda_Q\left(t|\bar{V}(T), Y\right) = \lim_{h\to 0} P\left(t \leqslant Q < t + h|\bar{V}(T), Y\right)/h$. We consider model A, in which we assume that $\lambda_Q\left(t|\bar{V}(T), Y\right)$ follows a stratified Cox proportional hazards model of the form

$$\lambda_Q\left(t|\bar{V}(T), Y\right) = \lambda_0\left(t|\bar{V}(t)\right)\exp\left(\alpha_0 Y\right), \tag{7.37}$$

where $\lambda_0\left(t|\bar{V}(t)\right)$ is an unrestricted positive function and α_0 is an unknown parameter. Equation (7.37) states that the hazard of dropout at time t depends in an arbitrary and unknown way on the observed past $\bar{V}(t)$ and also on the possibly unobserved future outcome Y through the term $\exp(\alpha_0 Y)$. A special case of (7.37), where $\alpha_0 = 0$, has been studied by Robins and Rotnitzky (1992). When $\alpha_0 = 0$, the dropout time is conditionally independent of Y given the observed $\bar{V}(t)$ and the data are referred to as coarsened at random, as defined by Heitjan and Rubin (1991). Here, we allow for nonzero α_0.

Unfortunately, we cannot estimate α_0 from the observed data because for any given α_0, we can find $\lambda_0(t|\bar{V}(t))$ to fit the marginal distribution F_O of the observed data. However, we further note that given α_0, $\lambda_0(t|\bar{V}(t))$ is uniquely determined by F_O. In practice, we fix α_0 to make inference about the mean of Y, say $\mu(\alpha_0)$, and then vary α_0 to perform the sensitive analysis. The model $A(\alpha_0)$, which differs from model A only in that α_0 is assumed known to the data analyst, is referred to as a ''nonparametric (just) identified'' model by Scharfstein et al. (1999).

The estimation of μ_0 under model $A(\alpha_0)$ might suffer from the curse of dimensionality. When the process $\bar{V}(t)$ is high dimensional, to produce a sufficiently good estimator of $\lambda_0(t|\bar{V}(t))$ and thus μ_0, it requires a large sample size, which is rare in practice. A process $\bar{V}(t)$ is usually regarded as high dimensional if (i) for each t, the vector $V(t)$ has two or more continuous components or many discrete components, or (ii) $\bar{V}(t)$ jumps at many different times. To avoid the curse of dimensionality, we consider model B, in which we assume that (7.37) holds but model $\lambda_0\left(t|\bar{V}(t)\right)$ as

$$\lambda_0\left(t|\bar{V}(t)\right) = \lambda_0(t)\exp\left(\gamma_0^{\top} W(t)\right), \tag{7.38}$$

where $\lambda_0(t)$ is an unspecified baseline hazard function, $W(t) = w(t, \bar{V}(t))$, $w(\cdot,\cdot)$ is a known function that maps $(t, \bar{V}(t))$ to $\mathbb{R}^q$, and γ_0 is a q-dimensional unknown parameter. Although $\lambda_0(t)$ is unrestricted, $\lambda_0\left(t|\bar{V}(t)\right)$ depends on $\bar{V}(t)$ only through $\gamma_0^{\top} W(t)$, which makes it possible not to use high-dimensional smoothing when estimating μ_0 under model $B(\alpha_0)$ (i.e., model B with α_0 assumed known).

Unlike model A, model B has no identifiability problem, and μ_0 and α_0 are often jointly identified because an additional functional form of $\lambda_0(t|\bar{V}(t))$ is imposed by (7.38). But if the dimension of $W(t)$ in (7.38) is chosen to be moderately large for robustness reason, the dependence of μ_0 and α_0 is generally weak and a large sample size is required for their joint estimation. So as done in model A, Scharfstein et al. (1999) recommended fixing α_0 in estimating μ_0 and varying α_0 in a sensitivity analysis.

7.6.2.2 Estimation in Model $A(\alpha_0)$

If there is no dropout, then μ_0 is estimated by the sample mean $n^{-1}\sum_{i=1}^{n} Y_i$. In the presence of dropout, although complete-case analysis usually leads to a biased estimator, by assigning appropriate weights to the non-dropout subjects, we can still obtain an unbiased estimator. Usually we choose the weight to be inverse of probability of non-dropout. Denote the probability of non-dropout by $\pi\left(\bar{V}(T), Y\right) = P\left(\Delta = 1|\bar{V}(T), Y\right)$. If $\lambda_0(t|\bar{V}(t))$, and thus by (7.37) $\pi\left(\bar{V}(T), Y\right)$ were known, we can easily develop the estimation of μ_0. If not, we need to estimate it. So to motivate the estimation method, in the following we start from the hypothetical case in which $\lambda_0(t|\bar{V}(t))$ were known.

Since Y is observed if the subject does not drop out before time T, we have $\pi\left(\bar{V}(T), Y\right) = S\left(T|\bar{V}(T), Y\right)$. Here, $S\left(t|\bar{V}(T), Y\right)$ is the survival function of Q, and by (7.37) we have $S\left(t|\bar{V}(T), Y\right) = \exp\left(-\Lambda_0\left(t|\bar{V}(t)\right)\right)\exp\left(\alpha_0 Y\right)$ with $\Lambda_0\left(t|\bar{V}(t)\right) = \int_0^t \lambda_0\left(u|\bar{V}(u)\right)du$ the cumulative conditional baseline hazard. Since $\lambda_0(t|\bar{V}(t))$ is assumed known, an estimator of μ_0 can be obtained by a method similar to Horvitz–Thompson (Horvitz and Thompson, 1952) estimator, that is, $n^{-1}\sum_{i=1}^{n} \Delta_i Y_i/\pi(\bar{V}_i(T), Y_i)$. We call this estimator the inverse probability of the censoring weighted (IPCW) estimator, for the observed probability is characterized by a censoring probability. Note that the IPCW estimator can be obtained by defining the estimating equation

$$\sum_{i=1}^{n} \frac{\Delta_i}{\pi\left(\bar{V}_i(T), Y_i\right)}\left(Y_i - \mu\right) = 0$$

with respect to μ. To improve the robustness of the estimation of μ_0, we introduce the augmented IPCW estimating equation by adding an augmentation term to it, that is,

$$\sum_{i=1}^{n}\left[\frac{\Delta_i}{\pi\left(\bar{V}_i(T), Y_i\right)}\left(Y_i-\mu\right)+a\left(O_i;\mu,\Lambda_0;b\right)\right]=0, \tag{7.39}$$

where $b = b(\bar{v}(t), t; \mu)$ is a function specified by the data analyst and

$$\begin{aligned} a\left(O;\mu,\Lambda_0;b\right) &= (1-\Delta)\, b\left(\bar{V}(Q), Q;\mu\right) \\ &\quad - E\left[(1-\Delta)\, b\left(\bar{V}(Q), Q;\mu\right) | \bar{V}(T), Y\right]. \end{aligned} \tag{7.40}$$

The conditional expectation in (7.40) can be explicitly evaluated as

$$\int_0^T b\left(\bar{V}(t), t;\mu\right)\exp\left(-\Lambda_0\left(t|\bar{V}(t)\right)\right)\exp\left(\alpha_0 Y\right)\lambda_0\left(t|\bar{V}(t)\right)\exp\left(\alpha_0 Y\right)\mathrm{d}t.$$

Denote the solution of (7.39) by $\tilde{\mu}(b)$. If $b(\bar{v}(t), t; \mu) \equiv 0$, then $a \equiv 0$ and the AIPCW estimating equation reduces to the IPCW estimating equation. When $b(\bar{v}(t), t; \mu)$ is nonzero, we refer to $\tilde{\mu}(b)$ as an AIPCW estimator.

Let $h\left(O;\mu,\Lambda_0;b\right) = \Delta(Y-\mu)/\pi\left(\bar{V}(T), Y\right) + a\left(O;\mu,\Lambda_0;b\right)$. We have $E\left[h(O;\mu_0,\Lambda_0;b)\right] = 0$ for any function b. Then, the asymptotic normality of $\tilde{\mu}(b)$ can be obtained by standard Taylor series arguments. Specifically, $n^{1/2}(\tilde{\mu}(b) - \mu_0)$ is asymptotically normal with mean 0 and variance var $\left\{h\left(O;\mu_0,\Lambda_0;b\right)\right\}$. The asymptotic variance can be consistently estimated by $n^{-1}\sum_{i=1}^{n} h\left(O_i;\tilde{\mu}(b), \Lambda_0;b\right)^2$.

In practice, the baseline hazard $\lambda_0(t|\bar{V}(t))$ and thus the cumulative baseline hazard $\Lambda_0(t|\bar{V}(t))$ is unknown. As long as we can estimate $\Lambda_0(t|\bar{V}(t))$, we can calculate $h\left(O_i;\mu,\hat{\Lambda};b\right)$, and then obtain an estimator $\hat{\mu}(b)$ by solving

$$\sum_{i=1}^{n} h\left(O_i;\mu,\hat{\Lambda};b\right) = 0.$$

Now we discuss the estimation of $\Lambda_0(t|\bar{V}(t))$. Here, we only consider the special case in which $\bar{V}(t)$ is time independent so that $\bar{V}(t) \equiv V$ for all t. This is because nonparametric estimation of $\Lambda_0(t|\bar{V}(t))$ suffers from the curse of dimensionality when $V(t)$ has multiple continuous components or the process $\bar{V}(t)$ jumps at many different times.

Suppose V is discrete. If V contains continuous components, they can be discretized into a finite number of groups. We could estimate $\Lambda(t|V)$ separately for each value of V. If Y is always observed, then we can employ the Nelson–Aalen estimator with censoring times as the jump times (Andersen et al., 1993). That is, we could estimate $\Lambda_0(t|V=v)$ by

$$\tilde{\Lambda}(t|V=v) = \int_0^t \left(\frac{1}{n_v} \sum_{i=1}^{n} I(V_i = v) \exp(\alpha_0 Y_i) I(Q_i \geqslant u) \right)^{-1} \times \left(\frac{1}{n_v} \sum_{i=1}^{n} \mathrm{d}N_i^v(u) \right), \qquad (7.41)$$

where $N_i^v(u) = I(V_i = v, Q_i \le u, \Delta_i = 0)$ and $n_v = \sum_{i=1}^n I(V_i = v)$. Since Y is not always observed, the integrand of (7.41) cannot be calculated. Inspired by the Horvitz–Thompson estimator, we notice that if appropriate weights are assigned to the completely observed subjects, the resulting quantity is observable and has the same probability limit as the integrand of (7.41). By noting that

$$E\left[\Delta I(V=v) S(u|V=v,Y)/S(T|V=v,Y)|V=v,Y,Q \geqslant u\right] = 1,$$

it follows from a uniform law of large numbers that

$$\sup_{u\in[0,T]} \left| \frac{1}{n_v} \sum_{i=1}^{n} I(V_i = v) \exp(\alpha_0 Y_i) I(Q_i \geqslant u) - \frac{1}{n_v} \sum_{i=1}^{n} \frac{\Delta_i I(V_i = v) \exp(\alpha_0 Y_i) I(Q_i \geqslant u)}{\exp\{-\exp(\alpha_0 Y_i)[\Lambda_0(T|V_i = v) - \Lambda_0(u|V_i = v)]\}} \right| \xrightarrow{P} 0.$$

Although the latter quantity within the absolute value depends on $\Lambda_0(\cdot|V=v)$, we could estimate it by the counterpart in (7.41) and then obtain the following recursive estimator:

$$\hat{\Lambda}(t|V=v) = \int_0^t \left(\frac{1}{n_v} \sum_{i=1}^{n} \frac{\Delta_i I(V_i = v) \exp(\alpha_0 Y_i) I(Q_i \geqslant u)}{\exp\{-\exp(\alpha_0 Y_i)[\hat{\Lambda}(T|V_i = v) - \hat{\Lambda}(u|V_i = v)]\}} \right)^{-1} \times \left(\frac{1}{n_v} \sum_{i=1}^{n} \mathrm{d}N_i^v(u) \right).$$

It can be easily seen that $\hat{\Lambda}(t|V=v)$ is a step function with jumps at each of the unique censoring times in the group with $V=v$. Therefore, we can obtain its explicit solution. Let $0 < Q_{(1)}^v < \cdots < Q_{(k_v)}^v \le T$ denote the unique censoring times, and c_k^v the number of subjects who are censored at $Q_{(k)}^v$, $k = 1, \ldots, k_v$. Let $\hat{\lambda}_k^v = \hat{\Lambda}\left(Q_{(k)}^v|V=v\right) - \hat{\Lambda}\left(Q_{(k-1)}^v|V=v\right)$ for $k = 2, \ldots, k_v$ and $\hat{\lambda}_1^v = \hat{\Lambda}\left(Q_{(1)}^v|V=v\right)$. Then, $\hat{\Lambda}(t|V=v) =$

$\sum_{k=1}^{k_v} \hat{\lambda}_k^v I(Q_{(k)}^v \leq t)$. Note that

$$\begin{aligned}
&\hat{\Lambda}\left(Q_{(k_v)}^v | V = v\right) \\
&= \sum_{k=1}^{k_v-1} \left(\sum_{i=1}^{n} \frac{\Delta_i I\left(V_i = v\right) \exp\left(\alpha_0 Y_i\right) I\left(Q_i \geqslant Q_{(k)}^v\right)}{\exp\left\{-\exp\left(\alpha_0 Y_i\right) \sum_{j=k+1}^{k_v} \hat{\lambda}_j^v\right\}} \right)^{-1} c_k^v \\
&\quad + \left(\sum_{i=1}^{n} \Delta_i I\left(V_i = v\right) \exp\left(\alpha_0 Y_i\right) I\left(Q_i \geqslant Q_{(k_v)}^v\right) \right)^{-1} c_{k_v}^v \\
&= \hat{\Lambda}\left(Q_{(k_v-1)}^v | V = v\right) + \left(\sum_{i=1}^{n} \Delta_i I\left(V_i = v\right) \exp\left(\alpha_0 Y_i\right) I\left(Q_i \geqslant Q_{(k_v)}^v\right) \right)^{-1} c_{k_v}^v.
\end{aligned}$$

We have $\hat{\lambda}_{k_v}^v = \left\{ \sum_{i=1}^n \Delta_i I\left(V_i = v\right) \exp\left(\alpha_0 Y_i\right) I\left(Q_i \geqslant Q_{(k_v)}^v\right) \right\}^{-1} c_{k_v}^v$. Similarly, we can have that for $k = k_v - 1, \ldots, 1$,

$$\hat{\lambda}_k^v = \left(\sum_{i=1}^{n} \frac{\Delta_i I\left(V_i = v\right) \exp\left(\alpha_0 Y_i\right) I\left(Q_i \geqslant Q_{(k)}^v\right)}{\exp\left\{-\exp\left(\alpha_0 Y_i\right) \sum_{j=k+1}^{k_v} \hat{\lambda}_j^v\right\}} \right)^{-1} c_k^v.$$

Thus, we can sequentially compute $\hat{\lambda}_k^v$ for $k = k_v, \ldots, 1$ and then obtain $\hat{\Lambda}(\cdot|V = v)$. Under regularity conditions, $n^{1/2}\{\hat{\Lambda}(\cdot|V = v) - \Lambda_0(\cdot|V = v)\}$ converges to a Gaussian process. Hence, $\hat{S}(t|V = v, Y) = \exp\{-\hat{\Lambda}(t|V = v)\exp(\alpha_0 Y)\}$ would be a consistent estimator of the survival function of Q given $V = v$ and Y.

Under regularity conditions, $\hat{\mu}(b)$ is asymptotically normal. Now we discuss how to estimate the asymptotic variance of $\hat{\mu}(b)$. The asymptotic variance of $\tilde{\mu}(b)$ can be consistently estimated by $n^{-1} \sum_{i=1}^n h\left(O_i; \tilde{\mu}(b), \Lambda_0; b\right)^2$. Unfortunately, in general, the asymptotic variance of $\hat{\mu}(b)$ cannot be estimated correspondingly by $n^{-1} \sum_{i=1}^n h\left(O_i; \hat{\mu}(b), \hat{\Lambda}; b\right)^2$, because estimation of $\Lambda_0(t|V)$ contributes to the asymptotic variance of $\hat{\mu}(b)$. However, there exists a function b^* for which estimation of $\Lambda_0(t|V)$ has no contribution to the asymptotic variance of $\hat{\mu}(b^*)$, and in fact $\hat{\mu}(b)$ has the same asymptotic variance for all functions b (Bickel et al., 1993). So for any given function b, the asymptotic variance of $\hat{\mu}(b)$ is consistently estimated by

$$n^{-1} \sum_{i=1}^n h\left(O_i; \hat{\mu}(b), \hat{\Lambda}; b^*\right)^2. \tag{7.42}$$

Scharfstein et al. (1999) showed that there exists one and only one such function b^* given by

$$b^*(V, t; \mu) = b^*(V; \mu) = E\left\{(Y - \mu)\exp\left(\alpha_0 Y\right) | V\right\} / E\left\{\exp\left(\alpha_0 Y\right) | V\right\}.$$

The function $b^*(V; \mu)$ is unknown and must be estimated. The conditional expectations of $b^*(V; \mu)$ can be consistently estimated by IPCW estimation:

$$\hat{E}\left[\ell(Y) | V = v\right] = \frac{1}{n_v} \sum_{i=1}^{n} \frac{\Delta_i I\left(V_i = v\right) \ell\left(Y_i\right)}{\hat{\pi}\left(v, Y_i\right)},$$

where $\ell(Y) = (Y - \mu)\exp\left(\alpha_0 Y\right)$ or $\exp\left(\alpha_0 Y\right)$. Then we estimate $b^*(v; \mu)$ by $\hat{b}^*(V; \mu) = \hat{E}\left\{(Y - \mu)\exp\left(\alpha_0 Y\right) | V\right\} / \hat{E}\left\{\exp\left(\alpha_0 Y\right) | V\right\}$. Replacing b^* with $\hat{b}^*$ in (7.42), we obtain a consistent estimator of asymptotic variance of $\hat{\mu}(b)$ for any b. It is suggested that in practice we use $\hat{\mu}(\hat{b}^*)$ to estimate μ_0, which can be written in the closed form as

$$\hat{\mu}\left(\hat{b}^*\right) = \frac{1}{n} \sum_{i=1}^{n} \frac{\Delta_i}{\hat{\pi}\left(V_i, Y_i\right)} Y_i - \left(\frac{\Delta_i - \hat{\pi}\left(V_i, Y_i\right)}{\hat{\pi}\left(V_i, Y_i\right)}\right) \frac{\hat{E}\left\{Y \exp\left(\alpha_0 Y\right) | V_i\right\}}{\hat{E}\left\{\exp\left(\alpha_0 Y\right) | V_i\right\}}.$$

7.6.2.3 *Estimation in Model* $B(\alpha_0)$ In model $A(\alpha_0)$, $\lambda_0(t|\bar{V}(t))$ is modeled nonparametrically, while in model $B(\alpha_0)$, it is modeled by (7.38) with parameter γ_0. So for model $B(\alpha_0)$, we need to estimate $\psi_0 = (\mu_0, \gamma_0^{\top})'$, although we are interested only in μ_0. The idea of the estimation approach in model $B(\alpha_0)$ is analogous to that in model $A(\alpha_0)$; we will discuss (i) the estimating equation when α_0 is assumed known; (ii) how to estimate λ_0, and (iii) the asymptotic variance of the estimator of ψ_0 and its estimate.

Suppose the baseline hazard $\lambda_0(t)$ in (7.38) is known, and we denote the cumulative baseline hazard by $\Lambda_0(t) = \int_0^t \lambda_0(u)\,\mathrm{d}u$. Recall that $W(t)$ and γ_0 in (7.38) are q-dimensional. To construct $q + 1$ AIPCW estimating functions, we introduce $q + 1$ functions $\mathbf{b} = (b_1, \ldots, b_{q+1})$, where $b_j = b_j(\bar{v}(t), t; \psi)$ are arbitrary real-valued functions of $(\bar{v}(t), t; \psi)$ that play the same role as b does in model $A(\alpha_0)$. The survival function of Q with undetermined parameter γ is given by

$$S\left(t|\bar{V}(T), Y; \gamma\right) = \exp\left\{-\int_0^t \exp\left(\gamma^{\top} W(u) + \alpha_0 Y\right) \mathrm{d}\Lambda_0(u)\right\}.$$

Let $\pi\left(\bar{V}(T), Y; \gamma\right) = S\left(T|\bar{V}(T), Y; \gamma\right)$ denote the probability of censoring. Then, we can estimate ψ_0 by the solution $\tilde{\psi}(\mathbf{b})$ to

$$\sum_{i=1}^{n} \mathbf{h}\left(O_i; \psi, \Lambda_0; \mathbf{b}\right) = 0,$$

where $\mathbf{h}(O;\psi,\Lambda_0;\mathbf{b}) = (h_1(O;\psi,\Lambda_0;b_1),\ldots,h_{q+1}(O;\psi,\Lambda_0;b_{q+1}))'$,

$$h_1\left(O;\psi,\Lambda_0;b_1\right) = \frac{\Delta(Y-\mu)}{\pi\left(\bar{V}(T),Y;\gamma\right)} + a\left(O;\psi,\Lambda_0;b_1\right),$$

and for $j \geq 2$,

$$h_j\left(O;\psi,\Lambda_0;b_1\right) = a\left(O;\psi,\Lambda_0;b_j\right).$$

Here,

$$\begin{aligned} a\left(O;\psi,\Lambda_0;b_j\right) = {} & -\frac{\Delta}{\pi\left(\bar{V}(T),Y;\gamma\right)} E_\gamma\left[(1-\Delta)\,b_j\left(\bar{V}(Q),Q;\psi\right)|\bar{V}(T),Y\right] \\ & + (1-\Delta)\,b_j\left(\bar{V}(Q),Q;\psi\right), \end{aligned}$$

and $E_\gamma\left[\cdot|\bar{V}(T),Y\right]$ is the expectation with respect to the distribution $F\left(t|\bar{V}(T),Y;\gamma\right) = 1 - S\left(t|\bar{V}(T),Y;\gamma\right)$.

The asymptotic normality of $\tilde{\psi}(\mathbf{b})$ can be easily obtained by the fact that $E\left[a\left(O;\psi_0,\Lambda_0;b_j\right)\right] = 0$ and thus $E\left[\mathbf{h}\left(O;\psi_0,\Lambda_0;\mathbf{b}\right)\right] = 0$. Provided that $\tau(\mathbf{b}) = \partial E\left[h\left(O;\psi_0,\Lambda_0;\mathbf{b}\right)\right]/\partial\psi$ is invertible, the asymptotic variance of $\tilde{\psi}(b)$ is given by $\tau(\mathbf{b})^{-1}\operatorname{var}\left\{h\left(O;\psi_0,\Lambda_0;\mathbf{b}\right)\right\}\left(\tau(\mathbf{b})^{-1}\right)'$, which can be consistently estimated by

$$\tilde{\tau}(\mathbf{b})^{-1}\left[n^{-1}\sum_{i=1}^{n}\mathbf{h}\left(O_i;\tilde{\psi}(\mathbf{b}),\Lambda_0;\mathbf{b}\right)^{\otimes 2}\right]\left(\tilde{\tau}(\mathbf{b})^{-1}\right)', \tag{7.43}$$

with $\tilde{\tau}(\mathbf{b}) = n^{-1}\sum_{i=1}^{n}\partial\mathbf{h}\left(O_i;\tilde{\psi}(\mathbf{b}),\Lambda_0;\mathbf{b}\right)/\partial\psi$.

The estimation of $\Lambda_0(t)$ is quite similar to that in Section 7.6.2.2. If we could always observe Y, then the estimator is given by

$$\tilde{\Lambda}(t;\gamma) = \int_0^t\left(\frac{1}{n}\sum_{i=1}^{n}\exp\left\{\gamma^\top W_i(u) + \alpha_0 Y_i\right\} I\left(Q_i \geqslant u\right)\right)^{-1}\left(\frac{1}{n}\sum_{i=1}^{n}\mathrm{d}N_i(u)\right),$$

where $N_i(u) = I\left(Q_i \leqslant u, \Delta_i = 0\right)$. Because Y is not always observed, we can define the recursive estimator by

$$\begin{aligned} & \hat{\Lambda}(t;\gamma) \\ & = \int_0^t\left(\frac{1}{n}\sum_{i=1}^{n}\frac{\Delta_i\exp\left\{\gamma^\top W_i(u) + \alpha_0 Y_i\right\} I\left(Q_i \geqslant u\right)}{\exp\left\{-\int_u^T\exp\left\{\gamma^\top W_i(x) + \alpha_0 Y_i\right\}\mathrm{d}\hat{\Lambda}(x;\gamma)\right\}}\right)^{-1}\left(\frac{1}{n}\sum_{i=1}^{n}\mathrm{d}N_i(u)\right). \end{aligned}$$

The explicit solution for $\hat{\Lambda}(t;\gamma)$ can be easily obtained as in Section 7.6.2.2. Given $\hat{\Lambda}(t;\gamma)$, we can define the estimating equation

$$\sum_{i=1}^{n} \mathbf{h}\left(O_i;\psi,\hat{\Lambda}(\cdot;\gamma);\mathbf{b}\right)=0$$

and the solution is denoted by $\hat{\psi}(\mathbf{b})$.

In general, the asymptotic variance of $\hat{\psi}(\mathbf{b})$ cannot be consistently estimated by an analog of (7.43) for a given function $\mathbf{b}$. However, as in Section 7.6.2.2, Scharfstein et al. (1999) showed that this is a case for a set of function $\mathbf{b}$, and a particular member $\mathbf{b}^*$ of this set is defined by giving an arbitrary $(q+1)$-dimensional function $\phi=\phi(\bar{V}(t),t)$ and solving the Volterra integral equation

$$\begin{aligned}&\mathbf{b}^*\left(\bar{V}(t),t;\psi_0\right)\\&=\phi\left(\bar{V}(t),t\right)-E\left\{S\left(t|\bar{V}(T),Y;\gamma_0\right)\exp\left(\gamma_0^\top W(t)+\alpha_0 Y\right)\right\}^{-1}q_{\mathbf{b}^*,\phi}\left(t;\psi_0\right),\end{aligned} \tag{7.44}$$

where

$$\begin{aligned}&q_{\mathbf{b},\phi}\left(t;\psi_0\right)\\&=E\left\{\phi\left(\bar{V}(t),t\right)S\left(t|\bar{V}(T),Y;\gamma_0\right)\exp\left(\gamma_0^\top W(t)+\alpha_0 Y\right)\right\}\\&\quad+\left\{\int_0^t E\left[\mathbf{b}\left(\bar{V}(u),u;\psi_0\right)\mathrm{d}F\left(u|\bar{V}(T),Y;\gamma_0\right)\exp\left(\gamma_0^\top W(t)+\alpha_0 Y\right)\right]\right\}\\&\quad-E\left\{\mathbf{e}_1\left(Y-\mu_0\right)\exp\left(\gamma_0^\top W(t)+\alpha_0 Y\right)\right\},\end{aligned}$$

and $\mathbf{e}_1$ is the $(q+1)$-dimensional vector with first component 1 and other components 0.

The last step is to estimate $\mathbf{b}^*$. After a function $\phi(\bar{V}(t),t)$ is selected, an estimator $\hat{\mathbf{b}}^*\left(\bar{V}(t),t;\psi\right)$, which is consistent at $\psi=\psi_0$ for the corresponding function $\mathbf{b}^*\left(\bar{V}(t),t;\psi\right)$, is obtained as follows. Let $Q_{(j)}$ be the jth ordered censoring time and define $Q_{(0)}=0$. The estimator $\hat{b}^*\left(\bar{V}(t),t;\psi\right)$ is obtained sequentially for $t\in(Q_{(k)},Q_{(k+1)}]$, $k=0,1,2,\ldots$. For $t\in(Q_{(k)},Q_{(k+1)}]$, let

$$\begin{aligned}&\hat{q}_{\mathbf{b},\phi}(t;\psi)\\&=\hat{E}_\gamma\left\{\phi\left(\bar{V}(t),t\right)\hat{S}\left(t|\bar{V}(T),Y;\gamma\right)\exp\left(\gamma^\top W(t)+\alpha_0 Y\right)\right\}\\&\quad+\sum_{j=1}^{k}\hat{E}_\gamma\left[\mathbf{b}\left(\bar{V}\left(Q_{(j)}\right),Q_{(j)};\psi\right)d\hat{F}\left(Q_{(j)}|\bar{V}(T),Y;\gamma\right)\exp\left(\gamma^\top W(t)+\alpha_0 Y\right)\right]\\&\quad-\hat{E}_\gamma\left\{\mathbf{e}_1\left(Y-\mu_0\right)\exp\left(\gamma^\top W(t)+\alpha_0 Y\right)\right\}\end{aligned}$$

and $\sum_{j=1}^{0}=0$, where for any $Z=z\left(\bar{V}(T),Y\right)$, $\hat{E}_\gamma(Z)$ is the IPCW estimate of $E_\gamma(Z)$, $n^{-1}\sum_{i=1}^{n}\Delta_i Z_i/\hat{\pi}\left(\bar{V}_i(T),Y_i;\gamma\right)$. Here, $\hat{\pi}$ is π with Λ_0 replaced by $\hat{\Lambda}$. Then, we can obtain that for $t \in (Q_{(k)}, Q_{(k+1)}]$,

$$\begin{aligned}&\hat{\mathbf{b}}^* \left(\bar{V}(t), t; \psi\right) \\ &= \phi\left(\bar{V}(t), t\right) - \hat{E}_\gamma\left\{\hat{S}\left(t|\bar{V}(T), Y; \gamma\right) \exp\left(\gamma^\top W(t) + \alpha_0 Y\right)\right\}^{-1} q_{\hat{\mathbf{b}}^*,\phi}(t; \psi).\end{aligned}$$

Therefore, the final estimator $\hat{\psi}(\hat{\mathbf{b}}^*)$ of ψ_0 has the asymptotic variance

$$\hat{\tau}\left(\hat{\mathbf{b}}^*\right)^{-1} \left[n^{-1} \sum_{i=1}^{n} \mathbf{h}\left(O_i; \hat{\psi}\left(\hat{\mathbf{b}}^*\right), \hat{\Lambda}\left(\cdot, \hat{\gamma}\left(\hat{\mathbf{b}}^*\right)\right); \hat{\mathbf{b}}^*\right)^{\otimes 2}\right] \left(\hat{\tau}\left(\hat{\mathbf{b}}^*\right)^{-1}\right)',$$

where $\hat{\tau}\left(\hat{\mathbf{b}}^*\right) = n^{-1} \sum_{i=1}^{n} \partial \mathbf{h}\left(O_i; \hat{\psi}\left(\hat{\mathbf{b}}^*\right), \hat{\Lambda}\left(\cdot, \hat{\gamma}\left(\hat{\mathbf{b}}^*\right)\right); \hat{\mathbf{b}}^*\right)/\partial\psi$.

7.6.2.4 Sensitivity Analysis We have introduced the methods of estimating the mean of Y under models $A(\alpha_0)$ and $B(\alpha_0)$ when the selection bias parameter α_0 is given. However, specifying a single value of α_0 is a difficult task in practice, so a sensitivity analysis is often desirable. A plausible range of α_0 can be determined based on one's scientific knowledge and a good understanding of the reasons of missing data. The population mean is then estimated with a series of different α_0 values, from which we can visualize how the estimator changes with different levels of nonignorable drop-out.

The sensitivity analysis provides us a range of results. If we desire to draw a final conclusion, we can impose a prior distribution on the selection bias parameter α_0 and summarize the results of the sensitivity analysis by Bayesian procedures. In this sense, Scharfstein et al. (1999) regarded a sensitivity analysis as useful ''preprocessing'' for any full Bayesian analysis that imposes prior distributions on α_0 and the other parameters.

CHAPTER 8

ANALYSIS OF RANDOMIZED CLINICAL TRIALS WITH NONCOMPLIANCE

8.1 OVERVIEW

In previous chapters, we introduced various methods for dealing with missing data. In this chapter, we apply some of these methods to make causal inference in randomized clinical trials with noncompliance and possibly missing outcome data.

A randomized clinical trial is the established gold standard for assessing the causal effect of a new intervention on patient outcomes. The validity of using a simple comparison of outcomes across intervention and control groups to estimate the causal effect of the intervention versus control depends on the assumption that the randomized experiment is implemented perfectly as designed. However, in practice, due to various reasons, many violations of the randomization protocol occur, and one of them is noncompliance, which occurs when the actual treatment a patient takes is different from the initially assigned treatment. Although on the surface it seems that the causal inference of randomized clinical data with noncompliance has nothing in common with the analysis of missing data, it turns out that techniques for dealing with missing data can also be adapted to estimate causal effects with randomized clinical data with noncompliance.

In practice, we encounter two types of noncompliance: (1) all-or-none noncompliance (e.g., either compliant or not) and (2) partial noncompliant. Since most of the available methods in the literature are for all-or-none compliance, in this chapter we focus on the problem of all-or-none compliance. For methods dealing with partial

Applied Missing Data Analysis in Health Sciences, First Edition. Xiao-Hua Zhou, Chuan Zhou, Danping Liu, and Xiaobo Ding.

noncompliance, we refer readers to the following references: Efron and Feldman (1991), Goetghebeur and Molenberghs (1996), Goetghebeur et al. (1998), and Jin and Rubin (2008).

For dealing with the problem of all-or-none compliance, there are four different classes of methods available: instrumental variables (IV) methods, moment-based methods, likelihood-based methods, and Bayesian methods. For example, Angrist et al. (1996) and Imbens and Rubin (1997b) proposed nonparametric methods for the causal analysis of studies with noncompliance using the idea of instrumental variables. Cuzick et al. (1997) developed likelihood methods for the causal analysis of studies with all-or-none compliance. Rubin (1974) and Imbens and Rubin (1997a) developed likelihood and Bayesian methods for the causal analysis of single follow-up studies without covariates; Hirano et al. (2000) extended their Bayesian methods to the single follow-up studies when the outcomes and compliance behavior might depend on some observed covariates.

In some clinical trials, subjects may also be missing their outcomes in addition to the problem of noncompliance. Several methods have also been developed to deal with both the problem of noncompliance and missing data. In general, two types of methods have been proposed, depending on how to model the missing data mechanism. One is based on the assumption that the missing data mechanism is latent ignorable (Frangakis and Rubin, 1999; Zhou and Li, 2006; Taylor and Zhou, 2009a, 2009b), and the other one is based on the nonignorable assumption (Chen et al., 2009).

This chapter is organized as follows. In Section 8.2, we introduce two real-world randomized clinical trials with noncompliance. In Section 8.3, we discuss some common but naive methods for dealing with noncompliance in the analysis of randomized clinical data with noncompliance. In Section 8.4, we introduce notation and define causal parameters of interest in the presence of noncompliance. In Section 8.5, we describe the nonparametric method with the use of instrumental variables for estimating causal effects. In Section 8.6, we introduce a moment-based method, which is similar to the instrumental variable method. In Section 8.7, we describe the maximum likelihood (ML) and Bayesian methods for binary outcomes. In Section 8.8, we discuss the problem of missing data in clinical trials with noncompliance. In Section 8.9, we analyze the two examples in Section 8.2 to illustrate the application of the proposed methods. In Section 8.10, we briefly discuss the multiple imputation method for dealing with noncompliance and MAR missing data and a nonparametric method for dealing with noncompliance and completely not missing at random (NMAR) data.

8.2 EXAMPLES

8.2.1 Randomized Trial on Vitamin A Supplement

Vitamin A is a crucial nutrient required for maintaining immune function, eye health, growth, and survival in human beings. Lack of vitamin A can lead to death. Sommer and Zeger (1997) reported a randomized clinical trial on the effectiveness

TABLE 8.1 Vitamin A Supplement Data

	$Z = 0, D = 0$	$Z = 1, D = 0$	$Z = 1, D = 1$
$Y = 0$	11514	2385	9663
$Y = 1$	74	34	12

of a vitamin A supplement in reducing mortality in children in rural Indonesia. In this trial, villages were randomized to either the treatment or the control group. All children in a village assigned to the treatment group would receive vitamin A supplements, and all children in a control village did not receive any vitamin A supplements. Due to various reasons, not all children in the treatment villages received a vitamin A supplement. The outcome of interest is the mortality rate within 12 months of the start of the study, that is, whether children survived at month 12 after the start of the study. In this example, we ignore any clustering within villages.

Let Z denote the treatment assignment for a patient; that is, $Z = 1$ if a patient was assigned to receive a vitamin A supplement, and $Z = 0$ if otherwise. Let D denote the actual treatment the patient received, that is, $D = 1$ if a patient received a vitamin A supplement, and $D = 0$ if a patient did not receive a vitamin A supplement. Let Y be a mortality binary indicator for a patient. Our observed data, taken from Imbens and Rubin (1997a), are summarized in Table 8.1. We are interested in estimating the effect of taking vitamin A supplement in reducing mortality rates.

8.2.2 Randomized Trial on Effectiveness of Flu Shot

From observational studies, it has been shown that the flu vaccination has improved outcomes in vaccinated patients. Hence, health officials in most countries recommend annual influenza vaccination for elderly persons and other people at high risk of influenza. However, no randomized controlled trials of the effects of influenza vaccination on pulmonary morbidity in high-risk adults have been published.

One possible reason for the lack of randomized trials is that widely accepted recommendations for vaccination raise ethical barriers against performing randomized controlled trials because it would require withholding vaccination from some subjects. One way around this impasse would be to perform a randomized trial of an intervention that increases the use of influenza vaccine in one group of patients without changing the use of influenza vaccine in another group. McDonald et al. (1992) used this idea to study influenza vaccine efficacy in reducing morbidity in high-risk adults, using a computer-generated reminder for flu shots.

The study was conducted over a 3-year period (1978–1980) in an academic primary care practice affiliated with a large urban public teaching hospital. Physicians in the practice were randomly assigned to either an the intervention or control group at baseline. Since physicians at the clinic each care for a fixed group of patients, their patients were similarly classified. During the study period, physicians in the intervention group received a computer-generated reminder when a patient with a

TABLE 8.2 Flu Shot Data

	$(Z, D) = (0, 0)$	(0,1)	(1,0)	(1,1)
$R = 1, Y = 0$	573	143	499	256
$R = 1, Y = 1$	49	16	47	20
$R = 0, Y = ?$	492	17	497	9

scheduled appointment was eligible for a flu shot. Since a patient whose physician received a computer reminder for the flu shot might not need to get a flu shot, some patients in the intervention group did not receive a flu shot. Similarly, some patients in the control group received their flu shots, even though their doctors did not receive a reminder for a flu shot. Hence, noncompliance exists in this data set. This data set has one additional problem of missing data for the outcome variable.

Let Z denote the treatment assignment for a patient, D denote the actual treatment the patient received, and Y be a binary indicator for flu-related hospitalization for a patient. Let R denote the missing outcome indicator. Our data are summarized in Table 8.2.

For the purpose of illustration, in this example we assume that outcomes of physician–patient pairs are independent, ignoring clustering of patients by physicians. One goal of this analysis was to estimate the causal effect of flu shot on flu-related hospitalization. See Frangakis et al. (2002) for an appropriate method that accounts for whether patients are clustered within their physicians.

8.3 SOME COMMON BUT NAIVE METHODS

When there exists noncompliance in a randomized clinical trial, three commonly used analytic methods are "intention to treat (ITT)," "per-protocol," and "as-treated" methods. An ITT analysis compares observed patient outcomes according to their initial treatment assignments and is the choice of many clinical trialists for the primary analysis of randomized trials (Friedman et al., 1985; Pocock, 1983). A "per-protocol" analysis compares observed outcomes for only those patients who appear to comply with the treatment assignment; that is, the analysis discards those patients who receive a treatment that is different from the assigned one and uses only those patients who receive the same treatment as assigned. An "as-treated" analysis compares observed outcomes according to the treatment actually received (Lee et al., 1991).

All the three methods have their limitations. Although the ITT method can provide a valid p-value of a test for the null hypothesis that both the treatment assignments and treatments actually received have no effects on patient outcomes (Rubin, 1998), it does not provide a valid estimate for the right question of interest, which is the causal effects of the treatment actually received on the outcome.

In contrast, both the "per-protocol" and "as-treated" methods attempt to address the right question of interest: causal effects of the treatment actually received on the outcome. However, both methods might result in a wrong answer to this question.

To illustrate the limitations of these three traditional methods, we use an example from Sheiner and Rubin (1995). In this example, we assume that a patient is randomly assigned to either a new drug (E) or an existing drug (C). We further assume that all C patients take the drug as assigned, whereas some E patients may not take the assigned new drug. Some patients who are assigned to the new drug E take it only for a short time and then switch to C due to its toxicity. We further assume that the new drug's toxicity is less tolerable to sick patients; therefore, noncompliers would more likely be those patients who are least likely to survive under either treatment.

The most commonly used ITT analysis compares outcomes between the treatment assignment groups. As the treatment assignment is randomized, the ITT analysis would give a reliable estimate of the causal effect of the treatment assignment on the outcome. However, the ITT results may give a biased estimate of the causal effect of the new drug on the outcome since the ITT analysis dilutes the effect of the new drug because of patients who are assigned to take the new drug but do not take it. Hence, the ITT likely underestimates the causal effect of the new drug on patient's outcome.

On the other hand, the per-protocol analysis discarded patients with poor prognosis and would overestimate the causal effect of the new drug on patient's outcome. The as-treated analysis ignores the initial random treatment assignment and compares patient's outcomes according to actual treatments patient received. In our simple example, some patients in the new drug group switched to the old drug group due to toxicity of the new drug. If poor prognosis patients tend to switch from the new drug to the old drug, the as-treated analysis would overestimate the causal effect of the new drug. If better prognosis patients tend to switch from the new drug to the old drug, the as-treated analysis would underestimate the causal effect of the new drug. An alternative method is needed to produce an unbiased or consistent estimate of the effect of drug (received) on outcome.

8.4 NOTATIONS, ASSUMPTIONS, AND CAUSAL DEFINITIONS

Let Z_i be the treatment assignment for the ith patient; Z_i is equal to 1 if the ith patient is assigned a treatment, and 0 for a control assignment.

Let $\mathbf{Z} = (Z_1, \ldots, Z_{\bar{N}})'$ be an $\bar{N}$-dimensional vector of treatment assignments for the $\bar{N}$ individuals in the population. Let $D_i(\mathbf{Z})$ be an indicator for the actual treatment the ith patient receives, given the treatment assignments of $\bar{N}$ individuals, $\mathbf{Z}$.

Let $\mathbf{D}(\mathbf{Z}) = (D_1(\mathbf{Z}), \ldots, D_N(\mathbf{Z}))'$. We also define the potential outcome $Y_i(\mathbf{Z}, \mathbf{D}(\mathbf{Z}))$, which is the potential outcome of the ith patient, given the treatment assignments and actual treatments received by all N individuals. In this notation, we allow the potential outcome of the ith individual to depend on the treatment assignments and actual treatments received by other individuals. To simplify this complication, we make the stable unit treatment value assumption (SUTVA), defined as follows:

Assumption 8.1 *(Stable Unit Treatment Value Assumption)*
(a) For any two treatment assignment vectors $\mathbf{Z}$ *and* $\mathbf{Z}'$, $D_i(\mathbf{Z}) = D_i(\mathbf{Z}')$, *if* $Z_i = Z_i'$.
(b) For any two treatment assignment vectors $\mathbf{Z}$ *and* $\mathbf{Z}'$, *and any two treatment vectors* $\mathbf{D}$ *and* $\mathbf{D}'$, *if* $Z_i = Z_i'$ *and* $D_i = D_i'$, *then* $Y_i(\mathbf{Z}, \mathbf{D}) = Y_i(\mathbf{Z}', \mathbf{D}')$.

This condition assumes that the potential outcome of each individual patient at a given time point does not depend on the treatment assignments or actual treatments received by other patients.

Under the SUTVA, we can simplify our notations as follows: $D_i(\mathbf{Z}) = D_i(Z_i)$ and $Y_i(\mathbf{Z}, \mathbf{D}) = Y_i(Z_i, D_i(Z_i))$. For example, for the flu shot data set, $Y_i(z, D_i(z)) = 1$ if the ith patient had a flu-related hospitalization, given that the treatment assignment of the ith patient is z.

The key concept in methods for dealing with noncompliance is to divide the patient population into subpopulations according to their potential compliance behavior. For the ith patient, let C_i be the compliance behavior, which is defined as follows:

$$C_i = \begin{cases} c \text{ (a complier)}, & \text{if } D_i(z) = z,\ z = 0, 1, \\ n \text{ (a never-taker)}, & \text{if } D_i(z) = 0,\ z = 0, 1, \\ a \text{ (an always-taker)}, & \text{if } D_i(z) = 1,\ z = 0, 1, \\ d \text{ (a defier)}, & \text{if } D_i(z) = 1 - z,\ z = 0, 1. \end{cases} \tag{8.1}$$

We observe the compliance behavior of a patient only partially, through the response to the actual treatment assignment the patient receives. If Z_i^{obs} denotes the actual treatment assignment of the ith individual, we can observe $D_i^{\text{obs}} = D_i(Z_i^{\text{obs}})$, but we cannot observe the response to the alternative treatment assignment, $D_i^{\text{mis}} = D_i(Z_i^{\text{mis}})$, where $Z_i^{\text{mis}} = 1 - Z_i^{\text{obs}}$. Let

$$\mathcal{C}(t) = \{i \mid C_i = t\}, \quad t = c, n, a, d.$$

Based on the compliance behavior, we can partition the population of patients into four subpopulations: $\mathcal{C}(c)$, $\mathcal{C}(n)$, $\mathcal{C}(a)$, and $\mathcal{C}(d)$.

We define the ITT causal effect of Z on the outcome Y to be

$$\text{ITT} = E(Y_i(1, D_i(1)) - Y_i(0, D_i(0))).$$

The population average ITT effect can be decomposed as

$$\text{ITT} = \sum_{t=\{c,n,a,d\}} P(C_i = t)\text{ITT}^t, \tag{8.2}$$

where

$$\text{ITT}^t = E(Y_i(1, D_i(1)) - Y_i(0, D_i(0)) \mid C_i = t)$$

is called the average ITT effect of the treatment assignment (Z) on the outcome (Y) in the subpopulation of individuals whose compliance behavior types belong to $\mathcal{C}(t)$.

Of the four subpopulation ITT effects, neither ITT^n nor ITT^a addresses causal effects of D (e.g., the receipt of flu shot) on Y because ITT^n compares the outcomes of two groups that did not receive the treatment, and ITT^a compares the outcomes of two groups that received the treatment. Only for the complier subpopulation and defier subpopulation, ITT^c and ITT^d compare the outcomes of those who received the treatment with the outcomes of those who did not receive the treatment.

From this discussion, we see that it is difficult to define causal effects of D on Y in the whole population without making additional assumptions. Hence, it may be more fruitful to assess causal effects of D on Y in the relevant subpopulations, the complier subpopulation and defier subpopulation. Since the existence of a defier group may not be realistic in practice, we focus on the subpopulation of compliers for the definition of the causal effects of D on Y. For the subpopulation of compliers, we define the average causal effect of D on Y as

$$\text{CACE} = \text{ITT}^c = E(Y_i(1,1) - Y_i(0,) \mid C_i = c),$$

called the complier average causal effect (CACE) (Imbens and Rubin, 1997a).

Next, we clarify information associated with each individual (Imbens and Rubin, 1997b). For the ith individual, we have defined five variables, Z_i, $D_i(0)$, $D_i(1)$, $Y_i(0, D_i(0))$, and $Y_i(1, D_i(1))$, as well as the sixth variable C_i, which is a function of $D_i(1)$ and $D_i(0)$. Among the five variables, three variables are observable: the treatment assignment, $Z_i^{\text{obs}} = Z_i$; the actual treatment received under the assigned treatment, $D_i^{\text{obs}} = D_i(Z_i^{\text{obs}})$; and the observed outcome under the assigned treatment and actual treatment received, $Y_i^{\text{obs}} = Y_i(Z_i^{\text{obs}}, D_i^{\text{obs}})$. The other two variables are missing: the treatment the ith individual would have received under the alternative treatment assignment to the actual treatment assignment, $D_i^{\text{mis}} = D_i(Z_i^{\text{mis}})$, where $Z_i^{\text{mis}} = 1 - Z_i^{\text{obs}}$; the outcome the ith individual would have observed under the alternative treatment assignment and corresponding actual treatment received, $Y_i^{\text{mis}} = Y_i(Z_i^{\text{mis}}, D_i^{\text{mis}})$. Of course, the sixth variable C_i is also missing.

To simplify the notation, for observed variables, we omit their subscripts. Hence, the observed data consist of (Z_i, D_i, Y_i), where $i = 1, \ldots, N$; here, N is the sample size. Let $\eta_{zt} = E(Y_i(z, D_i(z)) \mid Z_i = z, C_i = t)$, $\omega_t = P(C_i = t)$, and $\xi_z = P(Z_i = z)$, where $z = 0, 1$ and $t = n, a, c, d$. Without additional assumptions, these parameters are not identifiable, which means that different values of the parameters in the model can generate the probability distribution of the observed data. For a formal mathematical definition on model identifiability, see Lehmann and Casella (1998).

To make the model identifiable, we make three additional assumptions: strongly ignorable treatment assignment, monotonicity, and exclusion restriction. The strongly ignorable treatment assignment assumption is given as follows.

Assumption 8.2 *(Strongly Ignorable Treatment Assignment) Z_i and $(Y_i(0), Y_i(1), D_i(0), D_i(1))$ are independent.*

Note that a randomized trial satisfies this assumption.

The monotonicity assumption rules out the existence of defiers and is defined as follows.

Assumption 8.3 *(Monotonicity)*

$$D_i(1) \geq D_i(0), \textit{ for all } i = 1, 2, \ldots, \bar{N}.$$

In the flu shot example, the monotonicity assumption is plausible. It is unlikely physicians decide after receiving the reminder not to give flu shots to patients to whom they would have given a flu shot in the absence of the reminder. In general, for clinical trials with treatments that are believed to have some benefit, the monotonicity assumption is reasonable. Under the monotonicity assumption, $\omega_d = 0$. Hence, under the monotonicity assumption, the number of unknown parameters in the model is reduced by one.

The exclusion restriction assumption formalizes the notion that any effect of the treatment assignment on the outcome should be mediated by an effect of the treatment assignment on the actual treatment received. In the flu shot example, the exclusion restriction assumption means that any effect of the reminder on the outcome should be mediated by an effect of the reminder on the actual flu shot received. We only need to make the exclusion restriction assumption for never-takers and always-takers, formally defined as follows.

Assumption 8.4 *(Exclusion Restriction for Never-takers and Always-takers)*

$$f(Y_i(0, D_i(0)) \mid Z_i = 0, C_i = n) = f(Y_i(1, D_i(1)) \mid Z_i = 1, C_i = n),$$

$$f(Y_{ij}(0, D_i(0)) \mid Z_i = 0, C_i = a) = f(Y_{ij}(1, D_i(1)) \mid Z_i = 1, C_i = a).$$

Under the exclusion restriction assumption, we have $\eta_{zt} = \eta_t$, $t = n, a$, and have reduced the number of unknown parameters by two. Under the above four assumptions, it can be shown that the CACE is estimable (Angrist et al., 1996). Next, we will discuss the four methods for making inferences about CACE, including the method of instrumental variables, the moment-based method, the maximum likelihood method, and the Bayesian method.

8.5 METHOD OF INSTRUMENTAL VARIABLES

The method of instrumental variables is based on equation (8.2) and the exclusion assumption. From equation (8.2), we obtain

$$\text{ITT} = \sum_{t=\{c,n,a\}} \omega_t \text{ITT}^t.$$

From the exclusion restriction, $\mathrm{ITT}^n = \mathrm{ITT}^a = 0$. Therefore,

$$\mathrm{CACE} = \mathrm{ITT}^c = \frac{\mathrm{ITT}}{\omega_c}. \tag{8.3}$$

From this expression, we see that to estimate CACE, we need to estimate ω_c and ITT. Because of the strongly ignorable treatment assignment assumption and the exclusion restriction assumption, we have

$$\omega_n = P(C_i = n \mid Z_i = 1) = \frac{P(C_i = n, Z_i = 1, D_i = 0)}{P(Z_i = 1)} = \frac{P(Z_i = 1, D_i = 0)}{P(Z_i = 1)}.$$

From this formula, we obtain a moment-based estimator for ω_n as follows:

$$\widehat{\omega}_n = \frac{\sum_{i=1}^N Z_i(1 - D_i)}{\sum_{i=1}^N Z_i}.$$

Similarly, we obtain the following moment-based estimator for ω_a:

$$\widehat{\omega}_a = \frac{\sum_{i=1}^N (1 - Z_i) D_i}{\sum_{i=1}^N (1 - Z_i)}.$$

Since $\omega_c + \omega_a + \omega_n = 1$, we obtain the moment-based estimator for ω_c as follows:

$$\widehat{\omega}_c = 1 - \widehat{\omega}_n - \widehat{\omega}_a.$$

We can estimate ITT by the following consistent estimator:

$$\widehat{\mathrm{ITT}} = \frac{\sum_{i=1}^N Y_i Z_i}{\sum_{i=1}^N Z_i} - \frac{\sum_{i=1}^N Y_i(1 - Z_i)}{\sum_{i=1}^N (1 - Z_i)}.$$

Therefore, the IV estimator for CACE is given as follows:

$$\widehat{\mathrm{CACE}}_{\mathrm{IV}} = \frac{\widehat{\mathrm{ITT}}}{\widehat{\omega}_c}.$$

We can use the delta method to derive the asymptotic variance of $\widehat{\mathrm{CACE}}_{\mathrm{IV}}$ as follows (Imbens and Rubin, 1997a):

$$\mathrm{Var}(\widehat{\mathrm{CACE}}_{\mathrm{IV}}) = \frac{\mathrm{Var}(\widehat{\mathrm{ITT}})}{\omega_c^2} - 2\frac{\mathrm{ITT}}{\omega_c^3}\mathrm{Cov}(\widehat{\mathrm{ITT}}, \hat{\omega}_c) + \frac{\mathrm{ITT}^2}{\omega_c^4}\mathrm{Var}(\hat{\omega}_c).$$

Hence, we derive a variance estimator for the IV estimator as follows:

$$\widehat{\text{Var}}(\widehat{\text{CACE}}_{\text{IV}}) = \frac{\widehat{\text{Var}}(\widehat{\text{ITT}})}{\hat{\omega}_c^2} - 2\frac{\widehat{\text{ITT}}}{\hat{\omega}_c^3}\widehat{\text{Cov}}(\widehat{\text{ITT}}, \hat{\omega}_c) + \frac{\widehat{\text{ITT}}^2}{\hat{\omega}_c^4}\widehat{\text{Var}}(\hat{\omega}_c),$$

where

$$\widehat{\text{Var}}\left(\widehat{\text{ITT}}\right) = \frac{\sum_{i=1}^N Z_i\left(Y_i - \left[\left(\sum_{j=1}^N Y_j Z_j\right)/\left(\sum_{j=1}^N Z_j\right)\right]\right)^2}{\left(\sum_{i=1}^N Z_i\right)^2} + \frac{\sum_{i=1}^N (1-Z_i)\left\{Y_i - \left[\left(\sum_{j=1}^N Y_j (1-Z_j)\right)/\left(\sum_{j=1}^N (1-Z_j)\right)\right]\right\}^2}{\left\{\sum_{i=1}^N (1-Z_i)\right\}^2},$$

$$\widehat{\text{Var}}\left(\hat{\omega}_c\right) = \frac{\sum_{i=1}^N Z_i\left(D_i - \left[\left(\sum_{j=1}^N D_j Z_j\right)/\left(\sum_{j=1}^N Z_j\right)\right]\right)^2}{\left(\sum_{i=1}^N Z_i\right)^2} + \frac{\sum_{i=1}^N (1-Z_i)\left\{D_i - \left[\left(\sum_{j=1}^N D_j (1-Z_j)\right)/\left(\sum_{j=1}^N (1-Z_j)\right)\right]\right\}^2}{\left\{\sum_{i=1}^N (1-Z_i)\right\}^2},$$

and

$$\widehat{\text{Cov}}\left(\widehat{\text{ITT}}, \hat{\omega}_c\right) = \frac{\sum_{i=1}^N Z_i\left(Y_i - \left[\left(\sum_{j=1}^N Y_j Z_j\right)/\left(\sum_{j=1}^N Z_j\right)\right]\right)\left(D_i - \left[\left(\sum_{j=1}^N D_j Z_j\right)/\left(\sum_{j=1}^N Z_j\right)\right]\right)}{\left(\sum_{i=1}^N Z_i\right)^2}$$
$$+ \frac{\sum_{i=1}^N (1-Z_i)\left\{Y_i - \left[\left(\sum_{j=1}^N Y_j (1-Z_j)\right)/\left(\sum_{j=1}^N (1-Z_j)\right)\right]\right\}\left\{D_i - \left[\left(\sum_{j=1}^N D_j (1-Z_j)\right)/\left(\sum_{j=1}^N (1-Z_j)\right)\right]\right\}}{\left\{\sum_{i=1}^N (1-Z_i)\right\}^2}.$$

8.6 MOMENT-BASED METHOD

Since CACE $= \eta_{1c} - \eta_{0c}$, to estimate CACE, we need to estimate η_{0c} and η_{1c}. Under the monotonicity assumption, it is relatively easy to estimate η_n, η_a, and ω_t $(t = c, n, a)$ directly from the observed data.

We first show that η_{zc} $(z = 0, 1)$ can be written as functions of η_n, η_a, and ω_t as well as quantities that can be directly estimated from the observed data. Under the monotonicity assumption, we have $Z_i = 0$ and $C_i = c$ if and only if

$D_i(0) = 0, D_i(1) = 0$ or $D_i(0) = 0, D_i(1) = 1$. Therefore,

$$\begin{aligned}
&E(Y_i \mid Z_i = 0, D_i = 0) \\
&= E(Y_i \mid Z_i = 0, D_i = 0, C_i = n)P(C_i = n \mid Z_i = 0, D_i = 0) \\
&\quad + E(Y_i \mid Z_i = 0, D_i = 0, C_i = c)P(C_i = c \mid Z_i = 0, D_i = 0) \\
&= \eta_{0n} P(C_i = n \mid Z_i = 0, D_i = 0) + \eta_{0c} P(C_i = c \mid Z_i = 0, D_i = 0).
\end{aligned}$$

Note that

$$\begin{aligned}
&P(C_i = c \mid Z_i = 0, D_i = 0) \\
&= \frac{P(C_i = c, Z_i = 0, D_i = 0)}{\sum_{t=n,c} P(Z_i = 0, D_i = 0, C_i = t)} \\
&= \frac{P(D_i = 0 \mid C_i = c, Z_i = 0)P(C_i = c)P(Z_i = 0)}{\sum_{t=n,c} P(D_i = 0 \mid Z_i = 0, C_i = t)P(C_i = t)P(Z_i = 0)} \\
&= \frac{\omega_c}{\omega_c + \omega_n}.
\end{aligned}$$

Hence,

$$E(Y_i \mid Z_i = 0, D_i = 0) = \eta_n \frac{\omega_n}{\omega_c + \omega_n} + \eta_{0c} \frac{\omega_c}{\omega_c + \omega_n}.$$

Or,

$$\eta_{0c} = \frac{\omega_c + \omega_n}{\omega_c} \left[P(Y_i = 1 | Z_i = 0, D_i = 0) - \eta_n \frac{\omega_n}{\omega_c + \omega_n} \right]. \tag{8.4}$$

Similarly, we can obtain the following equation:

$$E(Y_i \mid Z_i = 1, D_i = 1) = \eta_a \frac{\omega_a}{\omega_c + \omega_a} + \eta_{1c} \frac{\omega_c}{\omega_c + \omega_a}.$$

Or,

$$\eta_{1c} = \frac{\omega_c + \omega_a}{\omega_c} \left[E(Y_i \mid Z_i = 1, D_i = 1) - \eta_a \frac{\omega_a}{\omega_c + \omega_a} \right]. \tag{8.5}$$

We observe that $E(Y_i \mid Z_i = 0, D_i = 1)$ and $E(Y_i \mid Z_i = 1, D_i = 1)$ can be estimated directly from the observed data. Therefore, to estimate η_{0c} and η_{1c}, we need to estimate η_n, η_a, and ω_t ($t = c, n, a$). We next derive moment-based estimators for η_n and η_a. Note that $Z_i = 1$ and $C_i = n$ if and only if $Z_i = 1$ and $D_i = 0$ because of the monotonicity assumption. Hence, we have

$$\eta_n = E(Y_i(1, D_i(1)) \mid Z_i = 1, C_i = n) = E(Y_i(1, D_i(1)) \mid Z_i = 1, D_i = 0).$$

Therefore, the moment-based estimator for η_n is

$$\widehat{\eta}_n = \frac{\sum_{i=1}^N Y_i Z_i(1 - D_i)}{\sum_{i=1}^N Z_i(1 - D_i)}.$$

Similarly, we can obtain a moment-based estimator for η_a as follows:

$$\widehat{\eta}_a = \frac{\sum_{i=1}^N Y_i(1 - Z_i)D_i}{\sum_{i=1}^N (1 - Z_i)D_i}.$$

We then derive moment-based estimators for ω_t $(t = c, n, a)$. Because of the strongly ignorable treatment assignment assumption, we have

$$\omega_n = P(C_i = n \mid Z_i = 1) = \frac{P(C_i = n, Z_i = 1, D_i = 0)}{P(Z_i = 1)} = \frac{P(Z_i = 1, D_i = 0)}{P(Z_i = 1)}.$$

From this formula, we obtain a moment-based estimator for ω_n as follows:

$$\widehat{\omega}_n = \frac{\sum_{i=1}^N Z_i(1 - D_i)}{\sum_{i=1}^N Z_i}.$$

Similarly, we obtain the following moment-based estimator for ω_a:

$$\widehat{\omega}_a = \frac{\sum_{i=1}^N (1 - Z_i)D_i}{\sum_{i=1}^N (1 - Z_i)}.$$

Since $\omega_c + \omega_a + \omega_n = 1$, we obtain a moment-based estimator for ω_c as follows:

$$\widehat{\omega}_c = 1 - \widehat{\omega}_n - \widehat{\omega}_a.$$

Using (8.4) and (8.5), we derive moment estimators for η_{0c} and η_{1c} as follows:

$$\widehat{\eta}_{0c} = \left[\frac{\sum_{i=1}^N Y_i(1 - Z_i)(1 - D_i)}{\sum_{i=1}^N (1 - Z_i)(1 - D_i)} - \widehat{\eta}_n \frac{\widehat{\omega}_n}{\widehat{\omega}_c + \widehat{\omega}_n} \right] \frac{\widehat{\omega}_c + \widehat{\omega}_n}{\widehat{\omega}_c} \qquad (8.6)$$

and

$$\widehat{\eta}_{1c} = \left[\frac{\sum_{i=1}^{N} Y_i Z_i D_i}{\sum_{i=1}^{N} Z_i D_i} - \widehat{\eta}_a \frac{\widehat{\omega}_a}{\widehat{\omega}_c + \widehat{\omega}_a} \right] \frac{\widehat{\omega}_c + \widehat{\omega}_a}{\widehat{\omega}_c}. \tag{8.7}$$

Hence, the moment-based estimator for CACE is given as follows:

$$\widehat{\text{CACE}}_m = \widehat{\eta}_{1c} - \widehat{\eta}_{0c},$$

where $\widehat{\eta}_{1c}$ and $\widehat{\eta}_{1c}$ are defined by (8.6) and (8.7), respectively. Similar to derivation of the variance for the IV estimator in Section 8.5, we can also use the delta method to derive the asymptotic variance of $\widehat{\text{CACE}}_m$.

8.7 MAXIMUM LIKELIHOOD AND BAYESIAN METHODS

If we are willing to assume a distribution for Y, we can also derive maximum likelihood estimators for CACE. In this section, we assume that Y is binary and that (Z, D, Y) follows a multinomial distribution. The log-likelihood function of the observed data (Y, Z, D) can then be written as follows:

$$\begin{aligned} l &= \sum_{z,d=0,1} \sum_{i \in S(z,d)} \log P(Y_i, Z_i = z, D_i = d) \\ &= \sum_{z,d=0,1} \sum_{i \in S(z,d)} \sum_{t=n,a,c,d} \log P(Y_i, Z_i = z, C_i = t, D_i = d) \\ &= \sum_{z,d=0,1} \sum_{i \in S(z,d)} \sum_{t=n,a,c,d} \log \eta_{zt}^{Y_i} (1 - \eta_{zt})^{1-Y_i} \omega_t \xi_z, \end{aligned}$$

where $S(z, d) = \{i : Z_i = z, D_i = d\}$, which is determined by $Z_i = z$ and $C_i = t$. This likelihood can be further simplified by writing out the likelihood contributions of $S(z, d)$.

- For $i \in S(0, 0)$,

$$\begin{aligned} & P(Y_i, Z_i = 0, D_i = 0) \\ = & P(Y_i, Z_i = 0, D_i = 0, C_i = c) + P(Y_i, Z_i = 0, D_i = 0, C_i = n) \\ = & P(Y_i \mid Z_i = 0, C_i = c) P(Z_i = 0) P(C_i = c) \\ & + P(Y_i \mid Z_i = 0, C_i = n) P(Z_i = 0) P(C_i = n) \\ & \times \left[\eta_{0c}^{Y_i} (1 - \eta_{0c})^{1-Y_i} \omega_c + \eta_n^{Y_i} (1 - \eta_n)^{1-Y_i} \omega_n \right] (1 - \xi). \end{aligned}$$

- For $i \in S(0,1)$,

$$
\begin{aligned}
P(Y_i, Z_i = 0, D_i = 1) &= P(Y_i, Z_i = 0, D_i = 1, C_i = a) \\
&= \eta_a^{Y_i}(1-\eta_a)^{1-Y_i}\omega_a(1-\xi).
\end{aligned}
$$

- For $i \in S(1,0)$,

$$
\begin{aligned}
P(Y_i, Z_i = 1, D_i = 0) &= P(Y_i, Z_i = 1, D_i = 0, C_i = n) \\
&= \eta_n^{Y_i}(1-\eta_n)^{1-Y_i}\omega_n\xi.
\end{aligned}
$$

- For $i \in S(1,1)$,

$$
\begin{aligned}
&P(Y_i, Z_i = 1, D_i = 1) \\
&= P(Y_i, Z_i = 1, D_i = 1, C_i = c) + P(Y_i, Z_i = 1, D_i = 1, C_i = a) \\
&= P(Y_i \mid Z_i = 1, C_i = c)P(Z_i = 1)P(C_i = c) \\
&\quad + P(Y_i \mid Z_i = 1, C_i = a)P(Z_i = 1)P(C_i = a) \\
&= \left[\eta_{1c}^{Y_i}(1-\eta_{1c})^{1-Y_i}\omega_c + \eta_a^{Y_i}(1-\eta_a)^{1-Y_i}\omega_a\right]\xi.
\end{aligned}
$$

Therefore, the log-likelihood function is given as follows:

$$
\begin{aligned}
l(\theta) = {} & r_{00}\log\left[\eta_{0c}\omega_c + \eta_n\omega_n\right](1-\xi) \\
& + \left(N_{00} - r_{00}\right)\log\left[(1-\eta_{0c})\omega_c(1-\eta_n)\right](1-\xi) \\
& + r_{01}\log\left[\eta_a\omega(1-\xi)\right] + (N_{01} - r_{01})\log\left[(1-\eta_a)\omega(1-\xi)\right] \\
& + r_{10}\log\left[\eta_n\omega_n\xi\right] + (N_{10} - r_{10})\log\left[(1-\eta_n)\omega_n\xi\right] \\
& + r_{11}\log[\eta_{1c}\omega_c + \eta_a\omega_a]\xi \\
& + (N_{11} - r_{11})\log[(1-\eta_{1c})\omega_c + (1-\eta_a)\omega_a]\xi.
\end{aligned}
$$

Maximization of the above likelihood function would give us the ML estimate for $\theta = (\eta_{tz}, \omega_t, \xi)$.

Since the observed-data likelihood, given above, has a complicated form, involving a mixture structure over a large amount of missing data, directly maximizing the observed-data likelihood may present a computational challenge. To overcome this computational problem, we can use the expectation–maximization (EM) algorithm for computing the ML estimator $\hat{\theta}$ for θ by treating C_i as missing data because the complete-data likelihood has a simple form if C_i were known for all patients.

The complete data would include Y_i, C_i, Z_i, and D_i. The complete-data likelihood function is given by

$$L_c(\theta) = \prod_{i=1}^{N} P[Y_i, C_i, Z_i, D_i] = \prod_{i=1}^{N} P[Y_i|C_i, Z_i, D_i]P[D_i|C_i, Z_i]P[C_i]P[Z_i].$$

Let $x_{yzt} = \sum_{i=1}^{N} I_{[Y_i=y, Z_i=z, C_i=t)}$, the number of subjects with $Y_i = y$, $Z_i = z$, and $C_i = t$, which is unobserved. We obtain the complete-data log-likelihood as follows:

$$\begin{aligned} l_c(\theta) = {} & x_{10n} \log[\eta_n \omega_n (1-\xi)] + x_{00n} \log[(1-\eta_n)\omega_n(1-\xi)] \\ & + x_{10a} \log[\eta_a \omega_a (1-\xi)] + x_{00a} \log[(1-\eta_a)\omega_a(1-\xi)] \\ & + x_{10c} \log[\eta_{0c} \omega_c (1-\xi)] + x_{00c} \log[(1-\eta_{0c})\omega_c(1-\xi)] \\ & + x_{11n} \log[\eta_n \omega_n \xi] + x_{01n} \log[(1-\eta_n)\omega_n \xi] \\ & + x_{11a} \log[\eta_a \omega_a \xi] + x_{01a} \log[(1-\eta_a)\omega_a \xi] \\ & + x_{11c} \log[\eta_{1c} \omega_c \xi] + x_{01c} \log[(1-\eta_{1c})\omega_c \xi]. \end{aligned}$$

The observed data consist of $Y = (Y_1, \ldots, Y_N)$, $D = (D_1, \ldots, D_N)$, and $Z = (Z_1, \ldots, Z_N)$.

As described in Chapter 4, the EM algorithm consists of iterations between E steps and M steps until convergence. In the $(k+1)$th step of the E step, we take the expectation of the complete-data log-likelihood, given the observed data and the previous parameter estimate $\theta = \theta^{(k)}$ at tth step, $E(l_c(\theta) \mid Y, D, \theta = \theta^{(k)})$. It is easy to show that

$$\begin{aligned} & E(l_c(\theta) \mid Y, D, Z, \theta = \theta^{(k)}) \\ = {} & x_{10n}^{(k)} \log(\eta_n \omega_n (1-\xi)) + x_{00n}^{(k)} \log((1-\eta_n)\omega_n(1-\xi)) \\ & + x_{10a}^{(k)} \log(\eta_a \omega_n (1-\xi)) + x_{00a}^{(k)} \log((1-\eta_a)\omega_n(1-\xi)) \\ & + x_{10c}^{(k)} \log(\eta_{0c} \omega_c (1-\xi)) + x_{00c}^{(k)} \log((1-\eta_{0c})\omega_c(1-\xi)) \\ & + x_{11n}^{(k)} \log(\eta_n \omega_n \xi) + x_{01n}^{(k)} \log((1-\eta_n)\omega_n \xi) \\ & + x_{11a}^{(k)} \log(\eta_a \omega_n \xi) + x_{01a}^{(k)} \log((1-\eta_a)\omega_a \xi) \\ & + x_{11c}^{(k)} \log(\eta_{1c} \omega_c \xi) + x_{01c}^{(k)} \log((1-\eta_{1c})\omega_c \xi), \end{aligned}$$

where

$$\begin{aligned} x_{yzt}^{(k)} &= E(x_{yzt} \mid Y, D, Z, \theta = \theta^{(k)}) \\ &= \sum_{i=1}^{N} E(I_{[Y_i=y]} I_{[Z_i=z]} I_{[C_i=t]} \mid Y_i = y, Z_i = z, D_i = d, \theta = \theta^{(k)}) \\ &= \sum_{i=1}^{N} I_{[Y_i=y]} I_{[Z_i=z]} P(C_i = t \mid Y_i = y, Z_i = z, D_i = d, \theta = \theta^{(k)}), \end{aligned}$$

with $d = D_i(z)$.

The M step involves maximizing $E(l_c(\theta) \mid Y, D, \theta = \theta^{(k)})$ with respect to θ, which gives us the following ML updated estimates:

$$\widehat{\xi}^{(k+1)} = \frac{x_{11n}^{(k)} + x_{01n}^{(k)} + x_{11a}^{(k)} + x_{01a}^{(k)} + x_{11c}^{(k)} + x_{01c}^{(k)}}{N},$$

$$\widehat{\eta}_n^{(k+1)} = \frac{x_{10n}^{(k)} + x_{11n}^{(k)}}{x_{10n}^{(k)} + x_{00n}^{(k)} + x_{11n}^{(k)} + x_{01n}^{(k)}},$$

$$\widehat{\eta}_a^{(k+1)} = \frac{x_{10a}^{(k)} + x_{11a}^{(k)}}{x_{10a}^{(k)} + x_{00a}^{(k)} + x_{11a}^{(k)} + x_{01a}^{(k)}},$$

$$\widehat{\eta_{0c}}^{(k+1)} = \frac{x_{10c}^{(k)}}{x_{10c}^{(k)} + x_{00c}^{(k)}}, \quad \widehat{\eta_{1c}}^{(k+1)} = \frac{x_{11c}^{(k)}}{x_{11c}^{(k)} + x_{01c}^{(k)}},$$

and

$$\widehat{\omega}_t^{(k+1)} = \frac{\sum_{z,y=0,1} x_{yzt}^{(k)}}{N}.$$

We iterate between the E and M steps until $\widehat{\theta}^{(k)}$ converges, and the convergent value $\widehat{\theta}$ is the ML estimator of θ. The variance of the ML estimator can be obtained from the Fisher information matrix.

By assuming a prior distribution for θ, the Bayesian analysis makes inferences about θ, based on the posterior distribution of θ, which is defined as the conditional probability of θ, given the observed data. Since CACE is a function of θ, we can also compute the posterior distribution of CACE from the posterior distribution of θ.

It is worth noting that the analysis of a nonidentified model is less an issue with the Bayesian method as frequentist-based methods as long as a proper prior is used. This is because a proper prior always leads to a proper posterior distribution, even with a non-identified model for observed data. With a proper posterior distribution, all model parameters can be estimated using the posterior distribution (Eberly and Carlin, 2000). However, Gelfand and Sahu (1999) pointed out that priors that are too informative can dominate the inference and priors near to be improper would produce ill-behaved posteriors. Hence, a Bayesian method with a proper prior may be used to estimate CACE using fewer assumptions, for example, without assuming the exclusion restriction.

We use the data augmentation (DA) to find the posterior distribution of θ, given the observed data, $p(\theta \mid Z_{\text{obs}}, D_{\text{obs}}, Y_{\text{obs}})$, because the complete-data likelihood function of $(Z_{\text{obs}}, D_{\text{obs}}, Y_{\text{obs}}, C)$ has a much simpler form than the observed-data likelihood function.

Similar to the EM algorithm, we obtain the complete-data likelihood as follows:

$$
\begin{aligned}
L_c(\theta) = {} & \xi^{N_1}(1-\xi)^{N_0}[\eta_n\omega_n(1-\xi)]^{x_{10n}}[(1-\eta_n)\omega_n(1-\xi)]^{x_{00n}} \\
& \times [\eta_a\omega_a(1-\xi)]^{x_{10a}}[(1-\eta_a)\omega_a(1-\xi)]^{x_{00a}} \\
& \times [\eta_{0c}\omega_c(1-\xi)]^{x_{10c}}[(1-\eta_{0c})\omega_c(1-\xi)]^{x_{00c}} \\
& \times [\eta_n\omega_n\xi]^{x_{11n}}[(1-\eta_n)\omega_n\xi]^{x_{01n}} \\
& \times [\eta_a\omega_a\xi]^{x_{11a}}[(1-\eta_a)\omega_a\xi]^{x_{01a}} \\
& \times [\eta_{1c}\omega_c\xi]^{x_{11c}}[(1-\eta_{1c})\omega_c\xi]^{x_{01c}}.
\end{aligned}
$$

DA requires three steps: (1) an initial draw from the prior for θ, (2) a draw from the complete-data posterior function, and (3) a draw from the conditional predictive distribution of C_i given Y_i, D_i, and Z_i.

The posterior distribution can be sensitive to the choice of a prior distribution, because the observed-data likelihood has a mixture structure over a large amount of missing data.

We can choose a proper prior distribution with a simple conjugate form. The prior below corresponds to adding to the likelihood function 30 extra observations: there are 10 additional observations for each complier type; for each complier type, the 10 additional observations are split into 2.5 for each of the four combinations of the binary variables (Z_i, Y_i):

$$
p(\theta) \approx \prod_{t\in\{c,n,a\}} \prod_{z\in\{0,1\}} \prod_{y\in\{0,1\}} \left[\xi_z\omega_t\eta_{zt}^{y}(1-\eta_{zt})^{1-y}\right]^{2.5}.
$$

With prior independence of parameters, the complete-data posterior distribution of θ can be written as the product of five distributions: one for x_i,

$$
p(\xi \mid C, Z_{\text{obs}}, D_{\text{obs}}, Y_{\text{obs}}) \approx \xi^{N_1+15}(1-\xi)^{N_0+15},
$$

three for the compliance types,

$$
p(\omega \mid C, Z_{\text{obs}}, D_{\text{obs}}, Y_{\text{obs}}) \approx \prod_{t\in\{n,a,c\}} \omega_t^{10+x_{10t}+x_{00t}+x_{11t}+x_{01t}},
$$

and one for the outcomes,

$$p(\eta \mid C, Z_{\text{obs}}, D_{\text{obs}}, Y_{\text{obs}}) \approx$$
$$\eta_{0c}^{5+x_{10c}+x_{00c}} \eta_{1c}^{5+x_{11c}+x_{01c}} \eta_n^{10+x_{11n}+x_{01n}+x_{10n}+x_{00n}} \eta_a^{10+x_{11a}+x_{01a}+x_{10a}+x_{00a}}.$$

8.8 NONCOMPLIANCE AND MISSING OUTCOME DATA

In practice, some clinical trials may also have missing outcomes for some subjects, in addition to noncompliance. Many methods have been proposed for dealing with both noncompliance and missing data. To get key ideas across, in this section, we introduce moment and maximum likelihood methods for estimating CACE under the latent ignorability assumption for the missing data mechanism when the outcome variable is binary in a randomized trial (Frangakis and Rubin, 1999; Zhou and Li, 2006).

To make a parameter identifiable, we need to make additional assumptions about the missing data mechanism. Here, we are making the latent ignorability assumption and the compound exclusion assumption proposed by Frangakis and Rubin (1999).

- *Latent Ignorability:* Potential outcomes and associated potential nonresponse indicators are independent within each level of the latent compliance covariate, that is,

$$P[R_i(1), R_i(0)|C_i, Y_i(1, D_i(1)), Y_i(0, D_i(0))] = P[R_i(1), R_i(0)|C_i].$$

- *Compound Exclusion Restrictions for Never-Takers and Always-Takers:* For never-takers and always-takers, Z influences Y and R only through D.

Here, the compound exclusion restriction for never-takers embodies the twin assumptions that within the subpopulation of never-takers, the distributions of the two potential outcomes $P[Y(1,0) = 1|Z = 1, D = 0]$ and $P[Y(0,0) = 1|Z = 0, D = 0]$ are identical, and that the distributions of the two missing data indicators $P[R(1) = 1|Z = 1, D = 0]$ and $P[R(0) = 1|Z = 0, D = 0]$ are also identical. In other words, the treatment assignment does not affect the outcome or missing data distributions for never-takers. The compound exclusion restriction for always-takers, similarly, comprises the twin assumptions that within the subpopulation of always-takers, the treatment assignment does not affect the outcome or missing data distributions.

To derive estimators for CACE, we need additional notations. Let $\gamma_{zt} \equiv P[R_i = 1|Z_i = z, C_i = t]$ be the probability of observing outcome Y for subjects with $Z_i = z$ and $C_i = t$. Let us define

$$\eta_{zt} = P(Y_i(z, D_i(z)) = 1 \mid Z_i = z, C_i = t),$$

the conditional probability of outcome $Y_i(z, D_i(z))$, given the complier type and the treatment assignment. We further define

$$\psi_{tzd} \equiv P[C_i = t | Z_i = z, D_i = d],$$

the probability of complier type, given the observed treatment assignment and receipt.

Under the compound exclusion assumption, we have

$$\eta_{1n} = \eta_{0n} \equiv \eta_n \text{ and } \gamma_{1n} = \gamma_{0n} \equiv \gamma_n.$$

Furthermore, under our assumptions, many of the conditional probabilities of complier types, given the observed data, are known:

$$\psi_{n00} = P[C_i = n | Z_i = 0, D_i = 0],\ \psi_{n10} = 1,\ \psi_{n01} = 0,\ \psi_{n11} = 0,$$

$$\psi_{a00} = 0,\ \psi_{a10} = 0,\ \psi_{a01} = 1,\ \psi_{a11} = P[C_i = a | Z_i = 1, D_i = 1],$$

$$\psi_{c00} = P[C_i = c | Z_i = 0, D_i = 0] = 1 - \psi_{n00},\ \psi_{c10} = 0,\ \psi_{c01} = 0,$$

$$\psi_{c11} = P[C_i = c | Z_i = 1, D_i = 1] = 1 - \psi_{a11}.$$

Therefore, among ψ_{zdt}, only ψ_{n00} and ψ_{a11} are unknown parameters. To simplify notation, we denote the two parameters of interest: $\psi_{n00} = \psi_n$ and $\psi_{a11} = \psi_a$.

The marginal distributions of potential outcomes and the missing data mechanism are defined by 10 parameters: $\psi_n, \psi_a, \eta_n, \eta_a, \eta_{0c}, \eta_{1c}, \gamma_n, \gamma_a, \gamma_{0c}$, and γ_{1c}. Let θ be a vector of these parameters; that is, $\theta = (\psi_n, \psi_a, \eta_n, \eta_a, \eta_{0c}, \eta_{1c}, \gamma_n, \gamma_a, \gamma_{0c}, \gamma_{1c})$.

We first introduce some notation to describe data.

- Let $N_{zd} \equiv \sum_{i=1}^{N} I_{[R_i=1, Z_i=z, D_i=d]}$, which represents the total number of subjects in the sample who have observed outcomes and who assigned to the treatment assignment z and who actually received the treatment assignment d.
- Let $r_{zd} \equiv \sum_{i=1}^{N} I_{[R_i=1, Y_i=1, Z_i=z, D_i=d]}$, which is the total number of hospitalized subjects assigned to the treatment group z, receiving treatment d, and having the observed outcome.
- Let $M_{zd} \equiv \sum_{i=1}^{N} I_{[R_i=0, Z_i=z, D_i=d]}$, which is the total number of subjects in the sample who are missing the outcome and who assigned to the treatment assignment z and who actually received the treatment assignment d.

Next, we describe two estimation methods for θ: the moment-based method and the maximum likelihood method. We start with finding moment-based estimators for parameters in noncomplier subpopulations.

We first derive moment-based estimators for ψ_n and ψ_a. Under randomization, we expect $P[Z = 0|C = n] = P[Z = 1|C = n]$. Because never-takers, by definition, have $D_i = 0$, we expect $P[Z = 0, D = 0|C = n] = P[Z = 1, D = 0|C = n]$. Hence, the expected number of never-takers with $(Z_i = 0, D_i = 0)$ is the same as the number of never-takers with ($Z_i = 1, D_i = 0$). Similarly, the expected number of always-takers with ($Z_i = 1, D_i = 1$) is the same as the number of always-takers with ($Z_i = 0, D_i = 1$). Then, we obtain that

$$\psi_n = \frac{P(C = n, Z = 0, D = 0)}{P(Z = 0, D = 0)} = \frac{P(C = n, Z = 1, D = 0)}{P(Z = 0, D = 0)} = \frac{P(Z = 1, D = 0)}{P(Z = 0, D = 0)}$$

and

$$\psi_a = \frac{P(C = c, Z = 1, D = 1)}{P(Z = 1, D = 1)} = \frac{P(C = c, Z = 0, D = 1)}{P(Z = 1, D = 1)} = \frac{P(Z = 0, D = 1)}{P(Z = 1, D = 1)}.$$

Therefore, the moment-based estimators for ψ_n and ψ_a are given as follows:

$$\widehat{\psi_n} = \frac{N_{10} + M_{10}}{N_{00} + M_{00}}, \qquad \widehat{\psi_a} = \frac{N_{01} + M_{01}}{N_{11} + M_{11}}.$$

We next derive the moment-based estimators for η_n, η_a, γ_n, and γ_a. By the latent ignorability assumption, we have $P[Y = 1|Z = 1, C = n] = P[Y = 1|Z = 1, C = n, R = 1]$ and $P[Y = 1|Z = 0, C = a] = P[Y = 1|Z = 0, C = a, R = 1]$. Hence, we have

$$\eta_n = \frac{P(Y = 1, Z = 1, C = n, R = 1)}{P(Z = 1, C = n, R = 1)} = \frac{P[Y = 1, Z = 1, D = 0, R = 1]}{P[Z = 1, D = 0, R = 1]}$$

and

$$\eta_a = \frac{P(Y = 1, Z = 0, C = a, R = 1)}{P(Z = 0, C = n, R = 1)} = \frac{P[Y = 1, Z = 0, D = 1, R = 1]}{P[Z = 0, D = 1, R = 1]}.$$

It follows that the moment-based estimators for η_n and η_a are given by

$$\widehat{\eta_n} = \frac{r_{10}}{N_{10}} \text{ and } \widehat{\eta_a} = \frac{r_{01}}{N_{01}},$$

respectively.

Using the compound exclusion assumption, we have the following results:

$$\gamma_n = P(R = 1|Z = 1, C = n) = \frac{P(R = 1, Z = 1, D = 0)}{P(Z = 1, D = 0)}$$

and

$$\gamma_a = P(R = 1|Z = 0, C = a) = \frac{P(R = 1, Z = 0, D = 1)}{P(Z = 0, D = 1)}.$$

Hence, the moment-based estimators for γ_n and γ_a are given by

$$\widehat{\gamma_n} = \frac{N_{10}}{N_{10} + M_{10}} \text{ and } \widehat{\gamma_a} = \frac{N_{01}}{N_{01} + M_{01}},$$

respectively.

We then derive moment-based estimators for parameters in complier sub-populations. We first observe that we can express the directly estimable quantity $P(R = 1 \mid Z = z, D = d)$ as mixtures of parameters of the noncomplier population, γ_{zt} and ψ_{zdt}, as well as the ratio of $P(R = 1, Z = z, D = \pi_{zd})$ and $P(Z = z, D = \xi_{zd})$. That is,

$$\begin{aligned}\frac{\pi_{zd}}{\xi_{zd}} &= P[R = 1|Z = z, D = d] \\ &= \sum_{t\in\{n,a,c\}} P[R = 1|Z = z, D = d, C = t]P[C = t|Z = z, D = d] \\ &= \sum_{t\in\{n,a,c\}} \gamma_{zt}\psi_{zdt}.\end{aligned}$$

By taking $z = 0$ and 1 in the above expression, we obtain

$$\frac{\pi_{00}}{\xi_{00}} = \gamma_n\psi_n + \gamma_{0c}(1-\psi_n), \quad \frac{\pi_{11}}{\xi_{11}} = \gamma_a\psi_a + \gamma_{1c}(1-\psi_a). \tag{8.8}$$

Solving γ_{0c} and γ_{1c} in (8.8), we obtain

$$\gamma_{0c} = \frac{1}{1-\psi_n}\left(\frac{\pi_{00}}{\xi_{00}} - \gamma_n\psi_n\right)$$

and

$$\gamma_{1c} = 11 - \psi_a\left(\frac{\pi_{11}}{\xi_{11}} - \gamma_a\psi_a\right).$$

Since

$$\sum_{i=1}^{N} R_i I_{[Z_i=d, D_i=d]}/N \quad \text{and} \quad \sum_{i=1}^{N} I_{[Z_i=d, D_i=d]}/N$$

are unbiased estimators of π_{zd} and ξ_{zd}, respectively, the moment-based estimators for γ_{0c} and γ_{1c} are

$$\widehat{\gamma_{0c}} = \frac{1}{1-\widehat{\psi_n}}\left(\frac{\widehat{\pi_{00}}}{\widehat{\xi_{00}}} - \widehat{\gamma_n}\widehat{\psi_n}\right)$$

and

$$\widehat{\gamma_{1c}} = \frac{1}{1-\widehat{\psi_a}}\left(\frac{\widehat{\pi_{11}}}{\widehat{\xi_{11}}} - \widehat{\gamma_a}\widehat{\psi_a}\right),$$

respectively, where

$$\widehat{\pi_{zd}} = \sum_{i=1}^{N} R_i I_{[Z_i=d, D_i=d]}/N = N_{zd}/N$$

and

$$\widehat{\xi_{zd}} = \sum_{i=1}^{N} I_{[Z_i=d, D_i=d]}/N = (N_{zd} + M_{zd})/N.$$

Using the same idea as before, we express parameters for the complier population, η_{0c} and η_{1c}, as a function of estimable quantities. We can show that

$$\begin{aligned}\frac{\nu_{00}}{\pi_{00}} &= P(Y=1|Z=0, D=0, R=1)\\ &= P(Y=1|Z=0, D=0, R=1, C=n)P(C=n|Z=0, D=0, R=1)\\ &\quad + P(Y=1|Z=0, D=0, R=1, C=c)P(C=c|Z=0, D=0, R=1).\end{aligned}$$

By simple calculation, we can show that

$$P(C=n|Z=0, D=0, R=1) = \frac{\gamma_n \psi_n}{\gamma_n \psi_n + \gamma_{0c}(1-\psi_n)}$$

and

$$P(C=c|Z=0, D=0, R=1) = \frac{\gamma_{0c}(1-\psi_n)}{\gamma_n \psi_n + \gamma_{0c}(1-\psi_n)}.$$

Therefore,

$$\frac{\nu_{00}}{\pi_{00}} = \eta_n \cdot \frac{\gamma_n \psi_n}{\gamma_n \psi_n + \gamma_{0c}(1-\psi_n)} + \eta_{0c} \cdot \frac{\gamma_{0c}(1-\psi_n)}{\gamma_n \psi_n + \gamma_{0c}(1-\psi_n)}.$$

Similar to the expression for ν_{00}/π_{00}, we obtain

$$\frac{\nu_{11}}{\pi_{11}} = \eta_a \cdot \frac{\gamma_a \psi_a}{\gamma_a \psi_a + \gamma_{1c}(1-\psi_a)} + \eta_{1c} \cdot \frac{\gamma_{1c}(1-\psi_a)}{\gamma_a \psi_a + \gamma_{1c}(1-\psi_a)}.$$

By solving for η_{0c} and η_{1c} in the above equations, we obtain

$$\eta_{0c} = \frac{1}{\gamma_{0c}(1-\psi_n)} \left(\frac{\nu_{00}}{\pi_{00}} \cdot [\gamma_n \psi_n + \gamma_{0c}(1-\psi_n)] - \eta_n \gamma_n \psi_n \right)$$

and

$$\eta_{1c} = \frac{1}{\gamma_{1c}(1-\psi_a)} \left(\frac{v_{11}}{\pi_{11}} \cdot [\gamma_a \psi_a + \gamma_{1c}(1-\psi_a)] - \eta_a \gamma_a \psi_a \right).$$

Hence, we obtain the following moment-based estimators for η_{0c} and η_{1c}:

$$\widehat{\eta_{0c}} = \frac{1}{1-\widehat{\gamma_{0c}}\widehat{\psi}_n} \left\{ \left[\widehat{\gamma}_n \widehat{\psi_n} + \gamma_{0c}\widehat{(1-\widehat{\psi_n})} \right] \cdot \frac{r_{00}}{N_{00}} - \widehat{\eta}_n \widehat{\gamma}_n \widehat{\psi_n} \right\},$$

$$\widehat{\eta_{0c}} = \frac{1}{1-\widehat{\gamma_{1c}}\widehat{\psi}_a} \left\{ \left[\widehat{\gamma}_a \widehat{\psi_a} + \gamma_{1c}\widehat{(1-\widehat{\psi_a})} \right] \cdot \frac{r_{11}}{N_{11}} - \widehat{\eta}_a \widehat{\gamma}_a \widehat{\psi_a} \right\}.$$

Using the above results, we can also estimate the proportions of the three complier types, ω_n, ω_a, and ω_c, as follows:

$$\widehat{\omega_n} = 2 \cdot \frac{N_{10}+M_{10}}{N}, \quad \widehat{\omega_a} = 2 \cdot \frac{N_{01}+M_{01}}{N}, \quad \text{and} \quad \widehat{\omega_c} = 1 - \widehat{\omega_n} - \widehat{\omega_a},$$

where N is the total number of subjects in the sample. The resulting estimator for the local CACE is given as

$$\widehat{\text{CACE}} = \widehat{\eta_{1c}} - \widehat{\eta_{0c}}.$$

We apply the bootstrap method to compute the standard errors for the moment estimators.

One potential problem with the moment estimators is that they are not constrained to the range of the parameters they estimate. Next, we develop maximum likelihood estimators for estimating causal parameters.

It can be shown by simple calculation that the observed-data log-likelihood $l(\theta) \equiv \log L(\theta)$ is given by

$$\begin{aligned}
l(\theta) = {} & r_{00} \log[\eta_n \gamma_n \omega_n + \eta_{0c}\gamma_{0c}(1-\omega_n-\omega_a)] \\
& + (N_{00}-r_{00}) \log[(1-\eta_n)\gamma_n\omega_n + (1-\eta_{0c})\gamma_{0c}(1-\omega_n-\omega_a)] \\
& + r_{10} \log(\eta_n\gamma_n\omega_n) + (N_{10}-r_{10}) \log[(1-\eta_n)\gamma_n\omega_n] \\
& + r_{01} \log(\eta_a\gamma_a\omega_a) + (N_{01}-r_{01}) \log[(1-\eta_a)\gamma_a\omega_a] \\
& + r_{11} \log[\eta_a\gamma_a\omega_a + \eta_{1c}\gamma_{1c}(1-\omega_n-\omega_a)] \\
& + (N_{11}-r_{11}) \log[(1-\eta_a)\gamma_a\omega_a + (1-\eta_{1c})\gamma_{1c}(1-\omega_n-\omega_a)] \\
& + M_{00} \log[(1-\gamma_n)\omega_n + (1-\gamma_{0c})(1-\omega_n-\omega_a)] \\
& + M_{10} \log[(1-\gamma_n)\omega_n] + M_{01} \log[(1-\gamma_a)\omega_a] \\
& + M_{11} \log[(1-\gamma_a)\omega_a + (1-\gamma_{1c})(1-\omega_n-\omega_a)] \\
& + N \log\left(\frac{1}{2}\right). \qquad (8.9)
\end{aligned}$$

To obtain the maximum likelihood estimator $\hat{\theta}$ for θ, we can directly solve the score function of θ, which is the partial derivative of the log-likelihood with respect to θ, $S(\theta) = \partial l(\theta)/\partial\theta$. Using the Fisher information matrix, we obtain standard error estimates for these parameters.

The above direct ML estimation, based on the observed data, has one major drawback: the observed-data likelihood, given in equation (8.9), has a complicated form, involving a mixture structure over a large amount of missing data. To overcome this drawback, we propose to use the EM algorithm for computing the ML estimator $\hat{\theta}$ for θ by treating C_i as missing data because the conditional complete-data likelihood has a simple form if C_i were known for all patients.

In the E step of the EM algorithm, we compute the conditional expectation of the complete-data likelihood function, given the previous parameter estimate, denoted by $\theta^{(k)}$, and the observed data. In the M step, we maximize this expectation with respect to θ, typically by differentiation, to obtain the new parameter estimate $\theta^{(k+1)}$. We repeat the E and M steps until the process converges at step K, where $\left|\theta^{(K+1)} - \theta^{(K)}\right| < \epsilon$. Here, ϵ is a very small constant. The one major advantage of using the EM algorithm here is that the EM algorithm yields an explicit solution for $\theta^{(k+1)}$ in terms of $\theta^{(k)}$ and the observed data, which simplifies computation. Next, we give the details on the E and M steps.

Let x_{yrzt} be the number of subjects with $Y_i = y$, $R_i = r$, $Z_i = z$, and $C_i = t$, $1 \leq i \leq N$. Then, we obtain the following complete-data log-likelihood:

$$
\begin{aligned}
l_c(\theta) = {} & x_{110n}\log(\eta_n\gamma_n\omega_n) + x_{110c}\log(\eta_{0c}\gamma_{0c}(1-\omega_n-\omega_n)) \\
& + x_{010n}\log[(1-\eta_n)\gamma_n\omega_n] + x_{010c}\log[(1-\eta_{0c})\gamma_{0c}(1-\omega_n-\omega_a)] \\
& + x_{111n}\log(\eta_n\gamma_n\omega_n) + x_{011n}\log((1-\eta_n)\gamma_n\omega_n) + x_{110a}\log(\eta_a\gamma_a\omega_a) \\
& + x_{010a}\log[(1-\eta_a)\gamma_a\omega_a] + x_{111a}\log(\eta_a\gamma_a\omega_a) \\
& + x_{111c}\log[\eta_{1c}\gamma_{1c}(1-\omega_n-\omega_a)] + x_{011a}\log[(1-\eta_a)\gamma_a\omega_a] \\
& + x_{011c}\log[(1-\eta_{1c})\gamma_{1c}(1-\omega_n-\omega_a)] + x_{100n}\log[\eta_n(1-\gamma_n)\omega_n] \\
& + x_{100c}\log[\eta_{0c}(1-\gamma_{0c})(1-\omega_n-\omega_a)] + x_{000n}\log[(1-\eta_n)(1-\gamma_n)\omega_n] \\
& + x_{000c}\log[(1-\eta_{0c})(1-\gamma_{0c})(1-\omega_n-\omega_a)] + x_{101n}\log[\eta_n(1-\gamma_n)\omega_n] \\
& + x_{001n}\log[(1-\eta_n)(1-\gamma_n)\omega_n] + x_{100a}\log[\eta_a(1-\gamma_a)\omega_a] \\
& + x_{000a}\log[(1-\eta_a)(1-\gamma_a)\omega_a] + x_{101a}\log[\eta_a(1-\gamma_a)\omega_a] \\
& + x_{101c}\log[\eta_{1c}(1-\gamma_{1c})(1-\omega_n-\omega_a)] + x_{001a}\log[(1-\eta_a)(1-\gamma_a)\omega_a] \\
& + x_{001c}\log[(1-\eta_{1c})(1-\gamma_{1c})(1-\omega_n-\omega_a)].
\end{aligned}
$$

In the E step, we take the expectation of the complete-data log-likelihood, given the observed data and the previous parameter estimate $\theta = \theta^{(k)}$, and obtain the following results:

$$x_{110n}^{(k)} = \frac{\eta_n^{(k)}\gamma_n^{(k)}\omega_n^{(k)}}{\eta_n^{(k)}\gamma_n^{(k)}\omega_n^{(k)} + \eta_{0c}^{(k)}\gamma_{0c}^{(k)}(1 - \omega_n^{(k)} - \omega_a^{(k)})} \cdot r_{00},$$

$$x_{110c}^{(k)} = \frac{\eta_{0c}^{(k)}\gamma_{0c}^{(k)}(1 - \omega_n^{(k)}\omega_n^{(k)}}{\eta_n^{(k)}\gamma_n^{(k)}\omega_n^{(k)} + \eta_{0c}^{(k)}\gamma_{0c}^{(k)}(1 - \omega_n^{(k)} - \omega_a^{(k)})} \cdot r_{00},$$

$$x_{010n}^{(k)} = \frac{(1-\eta_n)^{(k)}\gamma_n^{(k)}\omega_n^{(k)}}{(1-\eta_n)^{(k)}\gamma_n^{(k)}\omega_n^{(k)} + (1-\eta_{0c})^{(k)}\gamma_{0c}^{(k)}(1 - \omega_n^{(k)} - \omega_a^{(k)})} \cdot (N_{00} - r_{00}),$$

$$x_{010c}^{(k)} = \frac{(1-\eta_{0c})^{(k)}\gamma_{0c}^{(k)}(1 - \omega_n^{(k)} - \omega_a^{(k)})}{(1-\eta_n)^{(k)}\gamma_n^{(k)}\omega_n^{(k)} + (1-\eta_{0c})^{(k)}\gamma_{0c}^{(k)}(1 - \omega_n^{(k)} - \omega_a^{(k)})} \cdot (N_{00} - r_{00}),$$

$$x_{111n}^{(k)} = r_{10},$$

$$x_{011n}^{(k)} = N_{10} - r_{10},$$

$$x_{110a}^{(k)} = r_{01},$$

$$x_{010a}^{(k)} = N_{01} - r_{01},$$

$$x_{111a}^{(k)} = \frac{\eta_a^{(k)}\gamma_a^{(k)}\omega_a^{(k)}}{\eta_a^{(k)}\gamma_a^{(k)}\omega_a^{(k)} + \eta_{1c}^{(k)}\gamma_{1c}^{(k)}(1 - \omega_n^{(k)}) - \omega_a^{(k)}} \cdot r_{11},$$

$$x_{111c}^{(k)} = \frac{\eta_{1c}^{(k)}\gamma_{1c}^{(k)}(1 - \omega_n^{(k)}) - \omega_a^{(k)}}{\eta_a^{(k)}\gamma_a^{(k)}\omega_a^{(k)} + \eta_{1c}^{(k)}\gamma_{1c}^{(k)}(1 - \omega_n^{(k)}) - \omega_a^{(k)}} \cdot r_{11},$$

$$x_{011a}^{(k)} = \frac{\eta_a^{(k)}\gamma_a^{(k)}\omega_a^{(k)}}{\eta_a^{(k)}\gamma_a^{(k)}\omega_a^{(k)} + \eta_{1c}^{(k)}\gamma_{1c}^{(k)}(1 - \omega_n^{(k)} - \omega_a^{(k)})} \cdot (N_{11} - r_{11}),$$

$$x_{011c}^{(k)} = \frac{\eta_{1c}^{(k)}\gamma_{1c}^{(k)}(1 - \omega_n^{(k)} - \omega_a^{(k)})}{\eta_a^{(k)}\gamma_a^{(k)}\omega_a^{(k)} + \eta_{1c}^{(k)}\gamma_{1c}^{(k)}(1 - \omega_n^{(k)} - \omega_a^{(k)})} \cdot (N_{11} - r_{11}),$$

$$x_{100n}^{(k)} = \frac{\eta_n^{(k)}(1 - \gamma_n^{(k)})\omega_n^{(k)}}{(1 - \gamma_n^{(k)})\omega_n^{(k)} + (1 - \gamma_{0c}^{(k)})(1 - \omega_n^{(k)} - \omega_a^{(k)})} \cdot M_{00},$$

$$x_{100c}^{(k)} = \frac{\eta_{0c}^{(k)}(1 - \gamma_{0c}^{(k)})(1 - \omega_n^{(k)} - \omega_a^{(k)})}{(1 - \gamma_n^{(k)})\omega_n^{(k)} + (1 - \gamma_{0c}^{(k)})(1 - \omega_n^{(k)} - \omega_a^{(k)})} \cdot M_{00},$$

$$x_{000n}^{(k)} = \frac{(1-\eta_n^{(k)})(1-\gamma_n^{(k)})\omega_n^{(k)}}{(1-\gamma_n^{(k)})\omega_n^{(k)} + (1-\gamma_{0c}^{(k)})(1-\omega_n^{(k)}-\omega_a^{(k)})} \cdot M_{00},$$

$$x_{000c}^{(k)} = \frac{(1-\eta_{0c}^{(k)})(1-\gamma_{0c}^{(k)})(1-\omega_n^{(k)}-\omega_a^{(k)})}{(1-\gamma_n^{(k)})\omega_n^{(k)} + (1-\gamma_{0c}^{(k)})(1-\omega_n^{(k)}-\omega_a^{(k)})} \cdot M_{00},$$

$$x_{101n}^{(k)} = \eta_n^{(k)} \cdot M_{10},$$

$$x_{001n}^{(k)} = (1-\eta_n^{(k)}) \cdot M_{10},$$

$$x_{100a}^{(k)} = \eta_a^{(k)} \cdot M_{01},$$

$$x_{000a}^{(k)} = (1-\eta_a^{(k)}) \cdot M_{01},$$

$$x_{101a}^{(k)} = \frac{\eta_a^{(k)}(1-\gamma_a^{(k)})\omega_a^{(k)}}{(1-\gamma_a^{(k)})\omega_a^{(k)} + (1-\gamma_{1c}^{(k)})(1-\omega_n^{(k)}-\omega_a^{(k)})} \cdot M_{11},$$

$$x_{101c}^{(k)} = \frac{\eta_{1c}^{(k)}(1-\gamma_{1c}^{(k)})(1-\omega_n^{(k)}-\omega_a^{(k)})}{(1-\gamma_a^{(k)})\omega_a^{(k)} + (1-\gamma_{1c}^{(k)})(1-\omega_n^{(k)}-\omega_a^{(k)})} \cdot M_{11},$$

$$x_{001a}^{(k)} = \frac{(1-\eta_a^{(k)})(1-\gamma_a^{(k)})\omega_a^{(k)}}{(1-\gamma_a^{(k)})\omega_a^{(k)} + (1-\gamma_{1c}^{(k)})(1-\omega_n^{(k)}-\omega_a^{(k)})} \cdot M_1,$$

$$x_{001c}^{(k)} = \frac{(1-\eta_{1c}^{(k)})(1-\gamma_{1c}^{(k)})(1-\omega_n^{(k)}-\omega_a^{(k)})}{(1-\gamma_a^{(k)})\omega_a^{(k)} + (1-\gamma_{1c}^{(k)})(1-\omega_n^{(k)}-\omega_a^{(k)})} \cdot M_{11}.$$

In the M step, we find the next iteration estimate $\theta^{(k+1)}$ by finding the roots of the partial derivatives of $E\left[l_c(\theta)|Y, D, Z, \theta = \theta^{(k)})\right]$ with respect to θ,

$$S_c(\theta) = \frac{\partial E\left[l_c(\theta)|\underline{x}, \theta = \theta^{(k)})\right]}{\partial \theta},$$

where $Y_{obs} = (R_1 Y_1, \ldots, R_N Y_N) = (Y_1, \ldots, Y_n)$, $D = (D_1, \ldots, D_n)$, and $Z = (Z_1, \ldots, Z_n)$ are observed data. We can show that $S_c(\theta)$ has the following elements:

$$\dot{l}_c(\omega_n) = \frac{x_{110n}^{(k)} + x_{010n}^{(k)} + x_{111n}^{(k)} + x_{011n}^{(k)} + x_{100n}^{(k)} + x_{000n}^{(k)} + x_{101n}^{(k)} + x_{001n}^{(k)}}{\omega_n}$$
$$- \frac{x_{110c}^{(k)} + x_{010c}^{(k)} + x_{111c}^{(k)} + x_{011c}^{(k)} + x_{100c}^{(k)} + x_{000c}^{(k)} + x_{101c}^{(k)} + x_{001c}^{(k)}}{1-\omega_n-\omega_a},$$

$$\dot{l}_c(\omega_a) = \frac{x_{110a}^{(k)} + x_{010a}^{(k)} + x_{111a}^{(k)} + x_{011a}^{(k)} + x_{100a}^{(k)} + x_{000a}^{(k)} + x_{101a}^{(k)} + x_{001a}^{(k)}}{\omega_a}$$

$$-\frac{x_{110c}^{(k)}+x_{010c}^{(k)}+x_{111c}^{(k)}+x_{011c}^{(k)}+x_{100c}^{(k)}+x_{000c}^{(k)}+x_{101c}^{(k)}+x_{001c}^{(k)}}{1-\omega_n-\omega_a},$$

$$\dot{l}_c(\gamma_n)=\frac{x_{110n}^{(k)}+x_{010n}^{(k)}+x_{111n}^{(k)}+x_{011n}^{(k)}}{\gamma_n}-\frac{x_{100n}^{(k)}+x_{000n}^{(k)}+x_{101n}^{(k)}+x_{001n}^{(k)}}{1-\gamma_n},$$

$$\dot{l}_c(\gamma_a)=\frac{x_{110a}^{(k)}+x_{010a}^{(k)}+x_{111a}^{(k)}+x_{011a}^{(k)}}{\gamma_a}-\frac{x_{100a}^{(k)}+x_{000a}^{(k)}+x_{101a}^{(k)}+x_{001a}^{(k)}}{1-\gamma_a},$$

$$\dot{l}_c(\gamma_{0c})=\frac{x_{110c}+x_{010c}}{\gamma_{0c}}-\frac{x_{100c}+x_{000c}}{1-\gamma_{0c}},$$

$$\dot{l}_c(\gamma_{1c})=\frac{x_{111c}+x_{011c}}{\gamma_{1c}}-\frac{x_{101c}+x_{001c}}{1-\gamma_{1c}},$$

$$\dot{l}_c(\eta_n)=\frac{x_{110n}^{(k)}+x_{111n}^{(k)}+x_{100n}^{(k)}+x_{101n}^{(k)}}{\eta_n}-\frac{x_{010n}^{(k)}+x_{011n}^{(k)}+x_{000n}^{(k)}+x_{001n}^{(k)}}{1-\eta_n},$$

$$\dot{l}_c(\eta_a)=\frac{x_{110a}^{(k)}+x_{111a}^{(k)}+x_{100a}^{(k)}+x_{101a}^{(k)}}{\eta_a}-\frac{x_{010a}^{(k)}+x_{011a}^{(k)}+x_{000a}^{(k)}+x_{001a}^{(k)}}{1-\eta_a},$$

$$\dot{l}_c(\eta_{0c})=\frac{x_{110c}+x_{100c}}{\eta_{0c}}-\frac{x_{010c}+x_{000c}}{1-\eta_{0c}},$$

$$\dot{l}_c(\eta_{1c})=\frac{x_{111c}+x_{101c}}{\eta_{1c}}-\frac{x_{011c}+x_{001c}}{1-\eta_{1c}}.$$

Solving the above functions would give us the estimates of $\omega_n^{(k+1)}$, $\omega_a^{(k+1)}$, $\gamma_n^{(k+1)}$, $\gamma_a^{(k+1)}$, $\gamma_{0c}^{(k+1)}$, $\gamma_{1c}^{(k+1)}$, $\eta_n^{(k+1)}$, $\eta_a^{(k+1)}$, $\eta_{0c}^{(k+1)}$, and $\eta_{1c}^{(k+1)}$. We can also obtain $\psi_n^{(k+1)}$ and $\psi_a^{(k+1)}$ by noting that $\psi_n=\omega_n/(1-\omega_a)$ and $\psi_a=\omega_a/(1-\omega_n)$.

8.9 ANALYSIS OF THE TWO EXAMPLES

We illustrate the methods introduced in the previous sections with the two examples, described in Section 8.2.

8.9.1 Analysis of Vitamin A Supplement Example

Returning to the vitamin A supplement example, we are interested in estimating the causal effect of taking the vitamin A supplement on mortality. Since the mortality rate for children who were assigned the intervention but did not take the vitamin A supplement is 0.014, much higher than the mortality rate of control children, 0.006, it seems like intervention children who did not take vitamin A supplement are more sick than the control children who also did not take vitamin A supplement. Hence, the per-protocol analysis, which discard all intervention children who did not take vitamin A supplement, would produce a biased estimate of the effect of taking vitamin A supplement versus not taking vitamin A supplement because the intervention group of children was healthier than the control group.

To obtain unbiased estimates, we need to use IV methods. In this example, since no subjects from the control group receive vitamin A supplements, we have $D_i(0) = 0$, but $D_i(1) = 0$ or 1, where D_i denotes the receipt of vitamin A supplement for subject i. Hence, monotonicity holds, and $\omega_a = \omega_d = 0$. The estimated IV estimate is 0.0032 with a 90% confidence interval of (0.0009, 0.0055) (Imbens and Rubin, 2014).

Applying the ML and Bayesian methods to this data set under the exclusion restriction assumption, we obtain the MLE for CACE as 0.0032 with the 90% confidence interval of (0.0012, 0.0051). The mean of posterior mean under the exclusion is 0.0031 and 90% posterior confidence interval is (0.0012, 0.0051) (Imbens and Rubin, 2014).

8.9.2 Flu Shot Example

In the flu shot example, described in Section 8.2.2, we have the problem of both missing data and noncompliance. We apply the moment-based and ML methods, which were introduced in Section 8.8 and can handle the problem of both missing data and noncompliance.

With the ML method, we obtain the estimated CACE of −0.009 with a standard error of 0.112. With the moment method, we obtain the estimated CACE of 0.009 with a standard error of 0.479. The corresponding 95% confidence intervals for CACE using the ML and moment methods are (−0.211, 0.229) and (−0.948, 0.930), respectively.

Although the moment and ML estimates of CACE are different, both give the same conclusion that influenza vaccination is not associated with reduced risk of hospitalization for respiratory illness.

8.10 OTHER METHODS FOR DEALING WITH BOTH NONCOMPLIANCE AND MISSING DATA

In the previous section, we introduced the moment-based and ML methods for dealing with the problem of both noncompliance and missing data. There are other methods available for dealing with this problem. Taylor and Zhou (2009b) further extended the results of Zhou and Li (2006) by developing the asymptotic theory for their moment estimator and proposed a method to assess sensitivity of the moment method for violation of the assumption of latent ignorability.

Taylor and Zhou (2009a) have also developed the multiple imputation (MI) method for CACE. As discussed in the previous chapters, the MI method has several advantages over the ML methods. An important advantage of the MI analysis (and the Bayesian analysis) over and above the likelihood methods is that neither the monotonicity nor the compound exclusion restriction, which have been used to develop and justify estimation procedures, is essential, and violations of these assumptions can be addressed. However, for both Bayesian and MI methods, identification is based entirely on a carefully thought-out (or mathematically tractable) prior distribution.

MI has statistical properties that closely match the optimality of maximum likelihood methods (although not exactly optimal since ML involves no simulation), with the principal advantage being that it can be used in almost any situation, whereas ML is much more restricted in its applications, perhaps requiring specially designed EM algorithms and sometimes difficult analytic or numerical integration on the log-likelihood for interval estimates.

Another important advantage of the MI analysis over the Bayesian and ML methods is that MI works in conjunction with standard complete-data methods and software so that after the imputations are generated, data analysts who are not professional statisticians can apply standard complete-case analysis to the multiply imputed data sets. And since the imputation phase is operationally distinct from subsequent analysis, given a set of m imputations, many different analyses can be performed, making it unnecessary to reimpute when a new analysis is considered. By using a Bayesian framework for imputation and a frequentist approach to examining the estimator, one is also able to take advantage of the Bayesian framework, for example, by relaxing certain assumptions while having the ability to make direct comparisons with other frequentist methods. MI is especially promising when standard complete-case analysis is difficult to modify analytically in the presence of missing data, as in the case of nonignorable missing data or multivariate outcomes.

The above-discussed methods may still yield biased estimators of the CACE if the missing data mechanism is a different type of *not missing at random* (NMAR) from latent ignorability. Often, the mechanism of missing outcome Y can depend on missing values of Y. For example, some subjects may drop out of a study because of patient's worsening condition, Y; that is, the outcome Y has a direct effect on the missing data mechanism, which is called completely nonignorable missing data.

Chen et al. (2009) showed that, under certain conditions, the parameters of interest were identifiable even under completely nonignorable missing data. They then derived their maximum likelihood and moment estimators and evaluated their finite sample properties in simulation studies in terms of bias, efficiency, and robustness. Their simulation studies show that the proposed estimators behave reasonably well for finite sample sizes.

APPENDIX 8.A: MULTIVARIATE DELTA METHOD

The delta method is a powerful tool for deriving asymptotic distribution of a function of vector of random variables. Suppose that $X_1, X_2, \ldots$ are a sequence of some k-dimensional random vectors, which satisfy the following condition:

$$n^b(X_n - a) \to Y \text{ in distribution,}$$

where Y is some k-dimensional random vector. Also, suppose that g is a mapping from $\mathcal{R}^k$ to $\mathcal{R}^l$ with the finite first derivative with respect to a, $\partial g(a)/\partial a$, with $a \in \mathcal{R}_k$. Then, the multivariate delta method states the following result:

$$n^b(g(X_n) - g(a)) \to \frac{\partial g(a)}{\partial a} Y \text{ in distribution.}$$

A proof of this result can be obtained by using the first-order multivariate Taylor expansion (DasGupta, 2008; Vaart, 2000).

BIBLIOGRAPHY

Afifi, A. and Elashoff, R. (1966) Missing observations in multivariate statistics I: review of the literature. *Journal of the American Statistical Association*, 61:595–604.

Albert, J. H. and Chib, S. (1993) Bayesian analysis of binary and polychotomous response data. *Journal of the American Statistical Association*, 88:669–679.

Allison, P. D. (2001) *Missing Data*, Sage, Thousand Oaks, CA.

Andersen, P. K. and Gill, R. D. (1982) Cox's regression model for counting processes: a large-sample study. *Annals of Statistics*, 10:1100–1120.

Andersen, P. K., Borgan, O., Gill, R. D., and Keiding, N. (1993) *Statistical Models Based on Counting Processes*, Springer.

Andrige, R. R. and Little, R. J. A. (2010) A review of hot deck imputation for survey non-response. *International Statistical Review*, 78:40–64.

Angrist, J. D., Imbens, G. W., and Rubin, D.R. (1996) Identification of causal effects using instrumental variables. *Journal of the American Statistical Association*, 91: 444–455.

Bahadur, R. R. (1961) A representation of the joint distribution of responses to *n* dichotomous items. In: H. Solomon, editor, *Studies in Item Analysis and Prediction: Stanford Mathematical Studies in the Social Sciences VI*, Stanford University Press, Stanford, CA, pp. 158–168.

Barnard, J. and Rubin, D. B. (1999) Small sample degrees of freedom with multiple imputation. *Biometrika*, 86:948–955.

Basso, O., Pennell, M. L., Chen, A., and Longnecker, M. P. (2010) Mother's age at menarche and offspring size. *International Journal of Obesity*, 34:1766–1771.

Applied Missing Data Analysis in Health Sciences, First Edition. Xiao-Hua Zhou, Chuan Zhou, Danping Liu, and Xiaobo Ding.

Beekly, D. L., Ramos, E. M., van Belle, G., Deitrich, W., Clark, A. D., Jacka, M. E., Kukull, W. A., and the NIA-Alzheimer's Disease Canters (2004) The National Alzheimer's Coordinating Center (NACC) database: an Alzheimer disease database. *Alzheimer Disease and Associated Disorders*, 18:270–277.

Beekly, D. L., Ramos, E. M., Lee, W. W., Deitrich, W., Jacka, M. E., Wu, J., Hubbard, J. L., Koepsell, T. D., Morris, J. C., Kukull, W. A., and the NIA Alzheimer's Disease Centers (2007) The National Alzheimer's Coordinating Center (NACC) database: the uniform data set. *Alzheimer Disease and Associated Disorders*, 21:249–258.

Begun, J. M., Hall, W. J., Huang, W. M., and Wellner, J. A. (1983) Information and asymptotic efficiency in parametric–nonparametric models. *Annals of Statistics*, 11:432–452.

Beunckens, C., Sotto, C., and Molenberghs, G. (2008) A simulation study comparing weighted estimating equations with multiple imputation based estimating equations for longitudinal binary data. *Computational Statistics & Data Analysis*, 52:1533–1548.

Bickel, P. J., Klaasen, C. A., Ritov, Y., and Wellner, J. A. (1993) *Efficient and Adaptive Estimation for Semiparametric Models*, Johns Hopkins University Press.

Breslow, N. E. (2003) Whither PQL? UW Biostatistics Working Paper Series.

Carey, V., Zeger, S. L., and Diggle, P. J. (1993) Modelling multivariate binary data with alternating logistic regressions. *Biometrika*, 80:517–526.

Carpenter, J. and Kenward, M. G. (2013) *Multiple Imputation and Its Application*, *Statistics in Practice*, John Wiley & Sons, Ltd., Chichester, UK.

Cassel, C.-M., Särndal, C.-E., Wretman, J. H., Madow, W. G., and Olkin, I. (1983) Some uses of statistical models in connection with the nonresponse problem. In: Madow, W. G. and Olkin, I., editors, *Incomplete Data in Sample Surveys III. Proceedings of the Symposium on Incomplete Data*, Academic Press, New York.

Chen, B. and Zhou, X. H. (2011) Doubly robust estimates for binary longitudinal data analysis with missing response and missing covariates. *Biometrics*, 67:830–842.

Chen, B., Yi, G. Y., and Cook, R. J. (2010) Weighted generalized estimating functions for longitudinal response and covariate data that are missing at random. *Journal of the American Statistical Association*, 105:336–353.

Chen, C. F. (1985) On asymptotic normality of limiting density functions with Bayesian implications. *Journal of the Royal Statistical Society, Series B (Statistical Methodology)*, 97:540–546.

Chen, H., Geng, Z., and Zhou, X. H. (2009) Identifiability and estimation of causal effects in randomized trials with noncompliance and completely nonignorable missing data. *Biometrics*, 65:675–691.

Chen, H. Y. (2002) Double semiparametric method for missing covariates in Cox regression. *Journal of the American Statistical Association*, 97:565–576.

Chen, H. Y. (2004) Nonparametric and semiparametric models for missing covariates in parametric regression. *Journal of the American Statistical Association*, 99:1176–1189.

Chen, H. Y. and Little, R. J. (1999) Proportional hazards regression with missing covariates. *Journal of the American Statistical Association*, 94:896–908.

Chen, H. Y. and Little, R. J. (2001) A profile conditional likelihood approach for the semiparametric transformation regression model with missing covariates. *Lifetime Data Analysis*, 7:207–224.

Chen, M. H. and Ibrahim, J. G. (2001) Maximum likelihood methods for cure rate models with missing covariates. *Biometrics*, 57:43–52.

Chen, Z., Zhang, B., and Albert, P. S. (2011) A joint modeling approach to data with informative cluster size: robustness to the cluster size model. *Statistics in Medicine*, 30:1825–1836.

Cox, D. R. (1972) Regression models and life-tables (with discussion). *Journal of the Royal Statistical Society, Series B (Statistical Methodology)*, 34:187–220.

Cox, D. R. (1975) Partial likelihood. *Biometrika*, 62:269–276.

Curran, D., Molenberghs, G., Thijs, H., and Verbeke, G. (2004) Sensitivity analysis for pattern mixture models. *Journal of Biopharmaceutical Statistics*, 14:125–143.

Cuzick, J., Edward, R., and Segnan, N. (1997) Adjusting for non-compliance and contamination in randomized clinical trials. *Statistics in Medicine*, 16:1017–1029.

Daniels, J. and Hogan, J. W. (2008) *Missing Data in Longitudinal Studies*, Chapman & Hall/CRC.

DasGupta, A. (2008) *Asymptotic Theory of Statistics and Probability*, Springer, New York.

Demirtas, H., Arguelles, L. M., Chung, H., and Hedeker, D. (2007) On the performance of bias-reduction techniques for variance estimation in approximate Bayesian bootstrap imputation. *Computational Statistics and Data Analysis*, 51:4064–4068.

Dempster, A. P., Laird, N. M., and Rubin, D. B. (1977) Maximum likelihood from incomplete data via the EM algorithm (with discussion). *Journal of the Royal Statistical Society, Series B (Statistical Methodology)*, 39:1–38.

Diggle, P. J. and Kenward, M. G. (1994) Informative dropout in longitudinal data analysis (with discussion). *Journal of the Royal Statistical Society, Series C (Applied Statistics)*, 43:49–94.

Diggle, P. J., Heagerty, P., Liang, K. Y., and Zeger, S. L. (2002) *Analysis of Longitudinal Data*, Oxford University Press, Oxford.

Dunson, D. B., Chen, Z., and Harry, J. (2003) A Bayesian approach for joint modeling of cluster size and subunit-specific outcomes. *Biometrics*, 59:521–530.

Eberly, L. E. and Carlin, B. P. (2000) Identifiability and convergence issues for Markov chain Monte Carlo fitting of spatial models. *Statistics in Medicine*, 61:325–34.

Efron, B. and Feldman, D. (1991) Compliance as an explanatory variable in clinical trials. *Journal of the American Statistical Association*, 86:9–26.

Fay, R. E. (1996) Alternative paradigms for the analysis of imputed survey data. *Journal of the American Statistical Association*, 91:490–498.

Fitzmaurice, G. M. and Laird, N. M. (2000) Generalized linear mixture models for handling nonignorable dropouts in longitudinal studies. *Biostatistics*, 1:141–156.

Fitzmaurice, G. M., Molenberghs, G., and Lipsitz, S. R. (1995) Regression models for longitudinal binary responses with informative dropout. *Journal of the Royal Statistical Society, Series B (Statistical Methodology)*, 57:691–704.

Follmann, D. A. and Wu, M. C. (1995) An approximate generalized linear model with random effects for informative missing data. *Biometrics*, 51:151–168.

Fox, J. (2008) *Applied Regression Analysis and Generalized Linear Models*, Sage Publications, 2nd edition.

Frangakis, C. E. and Rubin, D. B. (1999) Addressing complications of intention-to-treat analysis in the combined presence of all-or-none treatment-noncompliance and subsequent missing outcomes. *Biometrika*, 86:365–379.

Frangakis, C. E., Rubin, D. B., and Zhou, X. H. (2002) Clustered encouragement designs with individual noncompliance: Bayesian inference with randomization, and application to advance directive forms (with discussion). *Biostatistics*, 3:147–164.

Friedman, L. M., Furberg, C. D., and DeMets, D. L. (1985) *Fundamentals of Clinical Trials*, PSG, Boston, MA.

Gelfand, A. E. and Sahu, S. K. (1999) Identifiability for improper priors and Gibbs sampling for generalized linear models. *Journal of the American Statistical Association*, 61:325–34.

Gelman, A. and Hill, J. (2006) Data Analysis Using Regression and Multilevel/Hierarchical Model, Cambridge University Press, New York, NY.

Gilks, W. R. and Wild, P. (1992) Adaptive rejection sampling for Gibbs sampling. *Applied Statistics*, 41:337–348.

Glynn, R. J., Laird, N. M., and Rubin, D. B. (1993) Multiple imputation in mixture models for nonignorable nonresponse with follow-ups. *Journal of the American Statistical Association*, 88:984–993.

Goetghebeur, E. and Molenberghs, G. (1996) Causal inference in a placebo-controlled clinical trial with binary outcome and ordered compliance. *Journal of the American Statistical Association*, 91:928–934.

Goetghebeur, E., Molenberghs, G., and Katz, J. (1998) Estimating the causal effect of compliance on binary outcome in randomized controlled trials. *Statistics in Medicine*, 17:341–355.

Goldstein, H. and Rasbash, J. (1996) Improved approximations for multilevel models with binary responses. *Journal of the Royal Statistical Society, Series A (Statistics in Society)*, 159:505–513.

Greenland, S. and Finkle, W. D. (1995) A critical look at methods for handling missing covariates in epidemiologic regression analyses. *American Journal of Epidemiology*, 142:1255–1264.

Heitjan, D. F. and Rubin, D. B. (1991) Ignorability and coarse data. *Annals of Statistics*, 19:2244–2253.

Hirano, K., Imbens, G. W., Rubin, D. B., and Zhou, X. H. (2000) Assessing the effect of an influenza vaccine in an encouragement design. *Biostatistics*, 1:1–20.

Hogan, J. W. and Laird, N. M. (1997) Mixture models for joint distribution of repeated measures and event times. *Statistics in Medicine*, 16:239–257.

Hogan, J. W., Roy, J., and Korkontzelou, C. (2004) Handling drop-out in longitudinal studies. *Statistics in Medicine*, 23:1455–1497.

Horton, N. J. and Laird, N. M. (1999) Maximum likelihood analysis of generalized linear models with missing covariates. *Statistical Methods in Medical Research*, 8:37–50.

Horvitz, D. G. and Thompson, D. J. (1952) A generalization of sampling without replacement from a finite universe. *Journal of the American Statistical Association*, 47: 663–685.

Huang, L., Chen, M. H., and Ibrahim, J. G. (2005) Bayesian analysis for generalized linear models with nonignorably missing covariates. *Biometrics*, 61:767–780.

Ibrahim, J. G. (1990) Incomplete data in generalized linear models. *Journal of the American Statistical Association*, 85:765–769.

Ibrahim, J. G., Lipsitz, S. R., and Chen, M. H. (1999) Missing covariates in generalized linear models when the missing data mechanism is nonignorable. *Journal of the Royal Statistical Society, Series B (Statistical Methodology)*, 61:173–190.

Ibrahim, J. G., Chen, M. H., Lipsitz, S. R., and Herring, A. H. (2005) Missing-data methods for generalized linear models. *Journal of the American Statistical Association*, 100: 332–346.

Imbens, G. and Rubin, D. B. (2014) *Causal Inference in Statistics and the Social Sciences*, Cambridge University Press, Cambridge, UK.

Imbens, G. W. and Rubin, D. B. (1997a) Bayesian inference for causal effects in randomized experiments with noncompliance. *Annals of Statistics*, 25:305–327.

Imbens, G. W. and Rubin, D. B. (1997b) Estimating outcome distributions for compliers in instrumental variables. *The Review of Economic Studies*, 64:555–574.

Jansen, I. and Molenberghs, G. (2010) Pattern-mixture models for categorical outcomes with non-monotone missingness. *Journal of Statistical Computation and Simulation*, 80:1279–1296.

Jin, H. and Rubin, D. B. (2008) Principal stratification for causal inference with extended partial compliance: application to Efron–Feldman data. *The Journal of the American Statistical Association*, 103:101–111.

Jones, M. P. (1996) Indicator and stratification methods for missing explanatory variables in multiple linear regression. *Journal of the American Statistical Association*, 91:222–230.

Langholz, B. and Goldstein, L. (1996) Risk set sampling in epidemiologic cohort studies. *Statistical Science*, 11:35–53.

Lavori, P. W., Brown, C. H., Duan, N. H., Gibbons, R. D., and Greenhouse, J. (2008) Missing data in longitudinal clinical trials part A: design and conceptual issues. *Psychiatric Annals*, 38:784–792.

Lee, Y. J., Ellenberg, J. H., Hirtz, D. G., and Nelson, K. B. (1991) Analysis of clinical trials by treatment actually received: is it really an option? *Statistics in Medicine*, 10: 1595–1605.

Lehmann, E. L. and Casella, G. (1998) *Theory of Point Estimation*, 2nd edition. Springer, New York.

Lei, H., Nahum-Shani, I., Lynch, K., Oslin, D., and Murphy, S. A. (2012) A ''SMART'' design for building individualized treatment sequences. *Annual Review of Clinical Psychology*, 8:21–48.

Li, K. H., Raghunathan, T. E., and Rubin, D. B. (1991) Large-sample significance levels from multiply imputed data using moment-based statistics and an F reference distribution. *Journal of the American Statistical Association*, 84:1065–1073.

Liang, K. Y. and Zeger, S. L. (1986) Longitudinal data analysis using generalized linear models. *Biometrika*, 73:13–22.

Lin, D. Y. and Ying, Z. (1993) Cox regression with incomplete covariate measurements. *Journal of the American Statistical Association*, 88:1341–1349.

Lin, E. H. B., Katon, W., Von Korff, M., Rutter, C., Simon, G. E., Oliver, M., Ciechanowski, P., Ludman, E. J., Bush, T., and Young, B. (2004) Relationship of depression and diabetes self-care, medication adherence, and preventive care. *Diabetes Care*, 27:2154–2160.

Lin, H., Liu, D., and Zhou, X. H. (2010) A correlated random-effects model for normal longitudinal data with nonignorable missingness. *Statistics in Medicine*, 29:236–247.

Lipsitz, S. R., Laird, N. M., and Harrington, D. P. (1991) Generalized estimating equations for correlated binary data: using the odds ratio as a measure of association. *Biometrika*, 78:153–160.

Little, R. J., Cohen, M. L., Dickersin, K., Emerson, S. S., Farrar, J. T., Neaton, J. D., Shih, W. J., Siegel, J. P., and Stern, H. (2012a) The design and conduct of clinical trials to limit missing data. *Statistics in Medicine*, 31:3433–3443.

Little, R. J., D'Aostino, R., Cohen, M. L., Dickersin, K., Emerson, S. S., Farrar, J. T., Frangakis, C., Hogan, J. W., Molenberghs, G., Murphy, S. A., Neaton, J. D., Rotnitzky, A., Scharfstein, D., Shih, W. J., Siegel, J. P., and Stern, H. (2012b) The prevention and treatment of missing data in clinical trials. *The New England Journal of Medicine*, 367:1355–1360.

Little, R. J. A. (1988) Missing-data adjustments in large surveys. *Journal of Business and Economic Statistics*, 6:287–301.

Little, R. J. A. (1993) Pattern-mixture models for multivariate incomplete data. *Journal of the American Statistical Association*, 88:125–134.

Little, R. J. A. (1995) Modelling the drop-out mechanism in repeated measures studies. *Journal of the American Statistical Association*, 90:1112–1121.

Little, R. J. A. and Rubin, D. B. (1987) *Statistical Analysis with Missing Data*, John Wiley & Sons, Inc., New York.

Little, R. J. A. and Rubin, D. B. (2002) *Statistical Analysis with Missing Data* 2nd edition, John Wiley & Sons, Ltd., Chichester, UK.

Little, R. J. A. and Wang, Y. (1996) Pattern-mixture models for multivariate incomplete data with covariates. *Biometrics*, 52:98–111.

Malakoff, D. (1999) Bayes offers a 'new' way to make sense of numbers. *Science*, 286:1460–1464.

McCullagh, P. and Nelder, J. A. (1989) *Generalized Linear Models*, 2nd edition, Chapman & Hall, London.

McDonald, C. J., Hui, S. L., and Tierney, W. M. (1992) Effects of computer reminders for influenza vaccination on morbidity during influenza epidemics. *M.D. Computing: Computers in Medical Practice*, 9:304–312.

McLachlan, G. J. and Krishnan, T. (1998) *The EM Algorithm and Extensions*, John Wiley & Sons, Inc., New York.

Meng, X. L. and Rubin, D. B. (1992) Performing likelihood ratio tests with multiply-imputed data sets. *Biometrika*, 79:103–111.

Molenberghs, G. and Kenward, M. G. (2007) *Missing Data in Clinical Studies*. John Wiley & Sons, Ltd. Chichester, UK.

Molenberghs, G., Michiels, B., Kenward, M. G., and Diggle, P. J. (1998) Monotone missing data and pattern-mixture models. *Statistica Neerlandica*, 52:153–161.

Monsell, S. E., Liu, D., Weintraub, S., and Kukull, W. A. (2012) Comparing measures of decline to dementia in amnestic MCI subjects in the National Alzheimer's Coordinating Center (NACC) uniform data set. *International Psychogeriatrics*, 24:1553–1560.

Mori, M., Woolson, R. F., and Woodworth, G. G. (1994) Slope estimation in the presence of informative right censoring: modeling the number of observations as a geometric random variable. *Biometrics*, 50:39–50.

National Research Council. (2010) *The Prevention and Treatment of Missing Data in Clinical Trials*, National Academies Press, Washington, DC.

Neyman, J. and Pearson, E. S. (1933) On the problem of the most efficient tests of statistical hypotheses. *Philosophical Transactions of the Royal Society A: Physical, Mathematical and Engineering Sciences*, 231:289–337.

Niswander, K. B. and Gordon, M. (1972) The women and their pregnancies: the collaborative perinatal study of the National Institute of Neurological Disease and Stroke. Technical report, US Government Printing Office, Washington, DC.

Ong, K. K., Northstone, K., Wells, J. C., Rubin, C., Ness, A. R., Golding, J. et al. Earlier mother's age at menarche predicts rapid infancy growth and childhood obesity. *PLoS Medicine*, 4:e132, 2007.

Paik, M. C. (1997) The generalized estimating equation approach when data are not missing completely at random. *Journal of the American Statistical Association*, 92:1320–1329.

Paik, M. C. and Tsai, W. Y. (1997) On using the Cox proportional hazards model with missing covariates. *Biometrika*, 84:579–593.

Patterson, H. D. and Thompson, R. (1971) Recovery of inter-block information when block sizes are unequal. *Biometrika*, 58:545–554.

Pepe, M. S. and Anderson, G. L. (1994) A cautionary note on inference for marginal regression models with longitudinal data and general correlated response data. *Communications in Statistics: Simulation and Computation*, 23:939–951.

Pinheiro, J. C. and Bates, D. M. (1995) Approximations to the log-likelihood function in the nonlinear mixed-effects model. *Journal of Computational and Graphical Statistics*, 4:12–35.

Pocock, S. J. (1983) *Clinical Trials: A Practical Approach*, John Wiley & Sons, Inc., New York.

Preisser, J. S., Lohman, K. K., and Rathouz, P. J. (2002) Performance of weighted estimating equations for longitudinal binary data with drop-outs missing at random. *Statistics in Medicine*, 21:3035–3054.

Prentice, R. L. (1988) Correlated binary regression with covariates specific to each binary observation. *Biometrics*, 44:1033–48.

Prentice, R. L. and Zhao, L. P. (1991) Estimating equations for parameters in means and covariances of multivariate discrete and continuous responses. *Biometrics*, 47:825–839.

Pugh, M., Robins, J., Lipsitz, S., and Harrington, D. (1993) Inference in the Cox proportional hazards model with missing covariate data. Technical report. Department of Biostatistics, Harvard School of Public Health, Boston.

Qi, L., Wang, C. Y., and Prentice, R. L. (2005) Weighted estimators for proportional hazards regression with missing covariates. *Journal of the American Statistical Association*, 100:1250–1263.

R Core Team. (2013) *R: A Language and Environment for Statistical Computing*. R Foundation for Statistical Computing, Vienna, Austria.

Rasbash, J., Steele, F., Browne, W.J., and Goldstein, H. (2012) *A User's Guide to MLwiN, v2.26*. Centre for Multilevel Modelling, University of Bristol.

Raudenbush, S. W., Yang, M. L., and Yosef, M. (2000) Maximum likelihood for generalized linear models with nested random effects via high-order, multivariate Laplace approximation. *Journal of Computational and Graphical Statistics*, 9:141–157.

Robins, J. M. (1992) Estimation of the time-dependent accelerated failure time model in the presence of confounding factors. *Biometrika*, 79:321–334.

Robins, J. M. (1994) Correcting for non-compliance in randomized trials using structural nested mean models. *Communications in Statistics*, 23:2379–2412.

Robins, J. M., Greenland, S. and Hu, F. C. (1999) Estimation of the causal effect of a time-varying exposure on the marginal mean of a repeated binary outcome. Journal of the American Statistical Association, 94, 687–700.

Robins, J. M. and Rotnitzky, A. (1992) Recovery of information and adjustment for dependent censoring using surrogate markers. In N. Jewek, K. Dietz, and V. Farewell, editors, *AIDS Epidemiology: Methodological Issues*, Birkhäuser, Boston, MA.

Robins, J. M. and Rotnitzky, A. (1995) Semiparametric efficiency in multivariate regression models with missing data. *Journal of the American Statistical Association*, 90:122–129.

Robins, J. M., Rotnitzky, A., and Zhao, L. P. (1994) Estimation of regression coefficients when some regressors are not always observed. *Journal of the American Statistical Association*, 89:846–866.

Robins, J. M., Rotnitzky, A., and Zhao, L. P. (1995) Analysis of semiparametric regression models for repeated outcomes in the presence of missing data. *Journal of the American Statistical Association*, 90:106–121.

Robinson, G. K. (1991) That BLUP is a good thing: the estimation of random effects. *Statistical Science*, 6:15–32.

Rodriguez, G. and Goldman, N. (1995) An assessment of estimation procedures for multilevel models with binary responses. *Journal of the Royal Statistical Society. Series A (Statistics in Society)*, 158:73–89.

Rotnitzky, A., Robins, J. M., and Scharfstein, D. O. (1998) Semiparametric regression for repeated outcomes with nonignorable nonresponse. *Journal of the American Statistical Association*, 93:1321–1339.

Rotnitzky, A., Scharfstein, D. O., Su, T. L., and Robins, J. M. (2001) Methods for conducting sensitivity analysis of trials with potentially nonignorable competing causes of censoring. *Biometrics*, 57:103–113.

Roy, J. (2003) Modeling longitudinal data with nonignorable dropouts using a latent dropout class model. *Biometrics*, 59:829–836.

Roy, J. and Daniels, M. J. (2008) A general class of pattern mixture models for nonignorable dropout with many possible dropout times. *Biometrics*, 64:538–545.

Rubin, D. B. (1974) Estimating causal effects of treatments in randomized and non-randomized studies. *Journal of Educational Psychology*, 66:688–701.

Rubin, D. B. (1987) *Multiple Imputation for Nonresponse in Surveys*, John Wiley & Sons, Inc., New York.

Rubin, D. B. (1998) More powerful randomization-based *p*-values in double-blinded trials with non-compliance. *Statistics in Medicine*, 17:371–385.

Rubin, D. B. and Schenker, N. (1986) Multiple imputation for interval estimation from simple random samples with ignorable nonresponse. *Journal of the American Statistical Association*, 81:366–374.

Rubin, D. B. and Schenker, N. (1991) Multiple imputation in health-care databases: an overview and some applications. *Statistics in Medicine*, 10:585–598.

Saha, C. and Jones, M. P. (2009) Bias in the last observation carried forward method under informative dropout. *Journal of Statistical Planning and Inference*, 139:246–255.

Schafer, J. L. (1997a) Imputation of missing covariates under a multivariate linear mixed model. Technical report, Department of Statistics, The Pennsylvania State University.

Schafer, J. L. (1997b) *Analysis of Incomplete Multivariate Data*, Chapman & Hall, London.

Schafer, J. L. (2013) *norm: Analysis of Multivariate Normal Datasets with Missing Values*. R package version 1.0-9.5 (http://CRAN.R-project.org/package=norm).

Schafer, J. L. and Graham, J. W. (2002) Missing data: our view of the state of the art. *Psychological Methods*, 7:147–177.

Schafer, J. L. and Schenker, N. (2000) Inference with imputed conditional means. *Journal of the American Statistical Association*, 95:144–154.

Scharfstein, D. O., Rotnitzky, A., and Robins, J. M. (1999) Adjusting for nonignorable drop-out using semiparametric nonresponse models. *Journal of the American Statistical Association*, 94:1096–1120.

Schenker, N. and Taylor, J. M. (1996) Partially parametric techniques for multiple imputation. *Computational Statistics and Data Analysis*, 22:425–446.

Seaman, S. and Copas, A. (2009) Doubly robust generalized estimating equations for longitudinal data. *Statistics in Medicine*, 28:937–955.

Sheiner, L. B. and Rubin, D. B. (1995) Intention-to-treat analysis and the goals of clinical trials. *Clinical Pharmacology & Therapeutics*, 57:6–15.

Shen, C. and Gao, S. (2007) A mixed-effects model for cognitive decline with non-monotone non-response from a two-phase longitudinal study of dementia. *Statistics in Medicine*, 26:409–425.

Siddique, J. and Belin, T. R. (2008a) Using an approximate Bayesian bootstrap to multiply impute nonignorable missing data. *Computational Statistical and Data Analysis*, 53:405–415.

Siddique, J. and Belin, T. R. (2008b) Multiple imputation using an iterative hot-deck with distance-based donor selection. *Statistics in Medicine*, 27:83–102.

Skinner, C. J. and D'Arrigo, J. (2011) Inverse probability weighting for clustered nonresponse. *Biometrika*, 98:953–966.

Some uses of statistical models in connection with the nonresponse problem, New York, 1983. Incomplete Data in Sample Surveys III. Symposium on Incomplete Data, Proceedings, Academic Press.

Sommer, A. and Zeger, S. L. (1997) Adjusting for non-compliance and contamination in randomized clinical trials. *Statistics in Medicine*, 16:1017–1029.

Stroup, T. S., McEvoy, J. P., Swattz, M. S., Byerly, M. J., Glick, I. D. Canive, J. M., McGee, M. F., Simpson, G. M., Stevens, M. C., and Lieberman, J. A. (2003) The National Institute of Mental Health Clinical Antipsychotic Trials of Intervention Effectiveness (CATIE) project: schizophrenia trials design and protocol development. *Schizophrenia Bulletin*, 29:15–31.

Tanner M. A. and Wong, W. H. (1987) The calculation of posterior distributions by data augmentation (with discussion). *Journal of the American Statistical Association*, 82:528–550.

Taylor, L. and Zhou, X. H. (2009a) Multiple imputation methods for treatment noncompliance and missing data. *Biometrics*, 65:88–95.

Taylor, L. and Zhou, X. H. (2009b) Relaxing latent ignorability in the ITT analysis of randomized studies with missing data and noncompliance. *Statistica Sinica*, 19:1203–1221.

Tsiatis, A. A. (1981) A large-sample study of Cox's regression model. *Annals of Statistics*, 9:93–108.

Tsiatis, A. A. (2006) *Semiparametric Theory and Missing Data*, Springer, New York.

Tsonaka, R., Verbeke, G., and Lesaffre, E. (2009) A semi-parametric shared parameter model to handle nonmonotone nonignorable missingness. *Biometrics*, 65:81–87.

Unützer, J., Katon, W., Callahan, C. M., Wallisms, Jr., J. W., Hunkeler, E., Harpole, L., Hoffing, M., Penna, R. D. D., Noël, P. H., Lin, E. H. B., Areán, P. A., Hegel, M. T., Tang, L., Belin, T. R., Oishi, S., and Langston, C. Collaborative care management of late-life depression in the primary care setting: a randomized controlled trial. *Journal of the American Medical Association*, 288:2836–2845, 2002.

van der Vaart, A. W. (2000) *Asymptotic Statistics*, Cambridge University Press, Cambridge, UK.

van Buuren, S. and Groothuis-Oudshoorn, K. (2011) MICE: multivariate imputation by chained equations in R. *Journal of Statistical Software*, 45:1–67.

van Buuren, S. (2013) *Flexible Imputation of Missing Data*, Chapman & Hall/CRC.

Verbeke, G. and Molenberghs, G. (2000) *Linear Mixed Models for Longitudinal data*, Springer, New York.

Wang, C. Y. and Chen, H. Y. (2001) Augmented inverse probability weighted estimator for Cox regression missing covariate regression. *Biometrics*, 57:414–419.

Wang, N. and Robins, J. M. (1998) Large sample inference in parametric multiple imputation. *Biometrika*, 85:935–948.

White, I. R. and Royston, P. (2009) Imputing missing covariate values for the Cox model. *Statistics in Medicine*, 28:1982–1998.

Wu, C. F. (1983) On the convergence properties of the EM algorithm. *Annals of Statistics*, 11:95–103.

Wu, M. C. and Carroll, R. J. (1988) Estimation and comparison of changes in the presence of informative right censoring by modelling the censoring process. *Biometrics*, 44: 175–188.

Xu, Q., Paik, M. C., Luo, X., and Tsai, W. Y. (2009) Reweighting estimators for Cox regression with missing covariates. *Journal of the American Statistical Association*, 104:1155–1167.

Yuan, Y. and Little, R. J. A. (2009) Mixed-effect hybrid models for longitudinal data with nonignorable dropout. *Biometrics*, 65:478–486.

Zhou, X. H. and Li, S. M. (2006) ITT analysis of randomized encouragement design studies with missing data. *Statistics in Medicine*, 25:2737–2761.

INDEX

Applied Missing Data Analysis in Health Sciences, First Edition. Xiao-Hua Zhou, Chuan Zhou, Danping Liu, and Xiaobo Ding.

WILEY SERIES IN STATISTICS IN PRACTICE

Advisory Editor, MARIAN SCOTT, *University of Glasgow, Scotland, UK*

Founding Editor, VIC BARNETT, *Nottingham Trent University, UK*

Human and Biological Sciences

Brown and Prescott · Applied Mixed Models in Medicine
Ellenberg, Fleming and DeMets · Data Monitoring Committees in Clinical Trials: A Practical Perspective
Lawson, Browne and Vidal Rodeiro · Disease Mapping With WinBUGS and MLwiN
Lui · Statistical Estimation of Epidemiological Risk
*Marubini and Valsecchi · Analysing Survival Data from Clinical Trials and Observation Studies
Parmigiani · Modeling in Medical Decision Making: A Bayesian Approach
Senn · Cross-over Trials in Clinical Research, *Second Edition*
Senn · Statistical Issues in Drug Development
Spiegelhalter, Abrams and Myles · Bayesian Approaches to Clinical Trials and Health-Care Evaluation
Turner · New Drug Development: Design, Methodology, and Analysis
Whitehead · Design and Analysis of Sequential Clinical Trials, *Revised Second Edition*
Whitehead · Meta-Analysis of Controlled Clinical Trials
Zhou, Zhou, Liu and Ding · Applied Missing Data Analysis in the Health Sciences

Earth and Environmental Sciences

Buck, Cavanagh and Litton · Bayesian Approach to Interpreting Archaeological Data
Cooke · Uncertainty Modeling in Dose Response: Bench Testing Environmental Toxicity
Gibbons, Bhaumik, and Aryal · Statistical Methods for Groundwater Monitoring, *Second Edition*
Glasbey and Horgan · Image Analysis in the Biological Sciences
Helsel · Nondetects and Data Analysis: Statistics for Censored Environmental Data
Helsel · Statistics for Censored Environmental Data Using Minitab® and R, *Second Edition*
McBride · Using Statistical Methods for Water Quality Management: Issues, Problems and Solutions
Ofungwu · Statistical Applications for Environmental Analysis and Risk Assessment
Webster and Oliver · Geostatistics for Environmental Scientists

Industry, Commerce and Finance

Aitken and Taroni · Statistics and the Evaluation of Evidence for Forensic Scientists, *Second Edition*
Brandimarte · Numerical Methods in Finance and Economics: A MATLAB-Based Introduction, *Second Edition*

*Now available in paperback.

Brandimarte and Zotteri · Introduction to Distribution Logistics
Chan and Wong · Simulation Techniques in Financial Risk Management
Jank · Statistical Methods in eCommerce Research
Jank and Shmueli · Modeling Online Auctions
Lehtonen and Pahkinen · Practical Methods for Design and Analysis of Complex Surveys, *Second Edition*
Lloyd · Data Driven Business Decisions
Ohser and Mücklich · Statistical Analysis of Microstructures in Materials Science
Rausand · Risk Assessment: Theory, Methods, and Applications